CW00767266

THE UNIVERSITY OF CHICAGO PRESS
CHICAGO, ILLINOIS

THE BAKER & TAYLOR COMPANY
NEW YORK

THE CAMBRIDGE UNIVERSITY PRESS
LONDON AND EDINBURGH

THE MARUZEN-KABUSHIKI-KAISHA
TOKYO, OSAKA, KYOTO, FUKUOKA, SENDAI

THE MISSION BOOK COMPANY
SHANGHAI

7210

A LABORATORY MANUAL

FOR

ELEMENTARY ZOÖLOGY

By

L. H. HYMAN

Department of Zoölogy, University of Chicago

THE UNIVERSITY OF CHICAGO PRESS
CHICAGO, ILLINOIS

COPYRIGHT 1919 BY
THE UNIVERSITY OF CHICAGO

All Rights Reserved

Published June, 1919

Composed and Printed By
The University of Chicago Press
Chicago, Illinois, U.S.A.

INTRODUCTORY NOTE

This laboratory manual was prepared for the class in elementary zoölogy in the University of Chicago, and has been used in this course for some time. In the three months' time allotted for the course, nearly all of the work given in sections I to XII, inclusive, is completed, with the exception of the muscular and skeletal systems of the frog. The remaining sections have been added for publication because they seem to the author to be essential for a complete course in general zoölogy. The directions have been written entirely from the material, although, of course, many textbooks have been consulted throughout. An attempt has been made, not merely to give explicit directions for the study of the material, but also to point out the relation of each section of the work to the general principles of biology and to make clear why each particular kind of material has been selected. The sections are given, naturally, in the order which the author thinks most logical, but they can be shifted at the will of the instructor. Our experience, extending over many years of trial of both methods, has shown that starting the course with the dissection of a complex animal is much more satisfactory than introducing the student at the start to the lower invertebrates, for the reason that the simplicity of the latter cannot be appreciated and understood except with reference to completely differentiated animals. Any vertebrate would serve the purpose, but the arthropods are in general unsuitable because they introduce the problem of heteronomous segmentation for which the beginner is not prepared. Many teachers will object that the instructions are too detailed and that it is better pedagogy to compel the student to work out things by himself; but the large amount of ground which must be covered in a relatively short time and the impossibility of providing sufficient laboratory assistants for the large classes with which we have to deal necessitate detailed directions. The student pursuing the modern college curriculum simply does not have time to carry out an original investigation on the anatomy of an animal; and unless he is provided with detailed directions, instructors and assistants are overburdened with the task of explaining to him what he is looking at, what to do next, etc. There is no reason why this extra work cannot be avoided by including these explanations in the manual.

GENERAL DIRECTIONS

1. Obtain from the cashier's office a biology breakage ticket ($5.00), and deposit it with the storekeeper in Room 10, Botany Building. Supplies needed for the laboratory are obtained from this storeroom, the cost is checked off on the breakage ticket, and a refund will be made for the unused portion of the ticket at the end of the year.

2. Obtain from the laboratory instructor a microscope card and present it at Room 2, Press Building, for a microscope. The microscope card will be issued only on presentation of the cashier's receipt showing that tuition has been paid. No charge is made for the use of the microscope.

3. Obtain at Room 10, Botany Building, after you have deposited the breakage ticket, the following supplies:
One fine scissors.
Two small forceps with straight points.
One medium-sized scalpel.
One probe.
Six slides.
Two teasing needles.
One dozen square cover glasses.
Three medicine droppers.
One tripod hand lens.
A paper of pins.
A towel.
A padlock.

If you already have some of these materials, do not, of course, purchase new ones.

4. Select a locker in the laboratory or have one assigned to you and place your padlock on the door. Keep your microscope, instruments, and drawing materials in this locker. If you want a clothes locker in the hall, buy a 50 cent key deposit ticket from the cashier's office, and present this for a key at Room 2, Press Building.

5. Purchase the following drawing materials (University Press or other bookstore):
Drawing paper, No. 6—a smooth, stiff paper is required.
Note paper, No. 6.
Notebook cover, No. 6.
Hard pencil, 6H.
Eraser.
Ruler.
Pad of sandpaper for sharpening the pencil.

vii

6. Provide yourself with a copy of Holmes's *Biology of the Frog*. Hegner's *An Introduction to Zoölogy* will be needed later.

7. Select a place in the laboratory and write your name on the table with a piece of chalk or paste a gummed label with your name on the table at the place chosen.

8. Present yourself with all of the foregoing materials ready for work at the beginning of the first laboratory period. Do not handicap yourself at the start by delaying to provide yourself with the necessary outfit. The microscope will not be needed for the first few days.

9. A box of slides will be issued to each student later in the course. Examine them to see that none are broken or damaged. These slides are to be returned at the end of the quarter, and a charge of 50 cents will be made for each broken or injured slide. Students are requested to handle the slides with care.

General Instructions Regarding Drawings

1. *All drawings must be made on good quality drawing paper with a hard pencil. As experience has shown that hard pencil drawings are the most satisfactory, no other kind will be accepted.*

2. All drawings must be made from the actual material, unless otherwise directed, and completed in the laboratory. The making of rough sketches in the laboratory to be completed elsewhere is unscientific, inaccurate, and not permissible. Drawings copied from textbooks will not be accepted.

3. The drawing should contain all structures mentioned in the outlines, as only those readily found are called for. In case you cannot find any structures mentioned, ask the assistants to help you. Put into the drawing only what you have actually seen, and in case you were unable to find certain things make a note to that effect under the drawing.

4. The prime requisite of a drawing is accuracy, i.e., it must resemble the actual specimen before the student as closely as possible. Drawings are not to be diagrammatized unless the outlines expressly direct to make them so. Next to accuracy, neatness and good arrangement on the page are desirable.

5. Make your drawings large enough to show clearly all details asked for; students tend to make drawings too small. The more details called for, the larger the drawing must be.

6. Always keep the pencil sharp by means of sandpaper.

7. Draw on one face of the page only, on the face which lies to your right hand with your notebook open.

8. Label fully. Label everything in the drawing regardless of whether the same structures have already been labeled in some previous drawing. With a ruler draw a straight line out from the object to be labeled and write or print the label so that it is parallel to the top and bottom edges of the page.

9. Do not write notes on the drawings. Write them on separate pages and insert them next to the drawings to which they refer.

10. Students who declare that they "cannot draw" will receive little sympathy. Anyone can make the simple line drawings required in this work. To make a drawing proceed as follows. First determine how large it is to be and select a proper space on the page. Then rule this space with very light vertical and horizontal guide lines so that your drawing will be symmetrical. With a ruler further reduce or enlarge the length or width of the actual object to fit the space selected. Then with very light lines make an outline of the object; then, constantly referring to the object, correct this with light lines until the proportions and details are as nearly like the object as you can possibly get them. Then erase the light lines until you can barely see them, and go over them making the final lines firm and clear. Every line upon the drawing must represent a structure actually present on the object. Avoid shading, color, crosshatching, etc.

Notes

1. Whenever desirable, notes are to accompany the drawings and are to be written on separate sheets which are to be inserted in the notebook adjacent to the sheet containing the drawings.

2. Avoid detailed notes containing material in the laboratory outlines.

3. The notes will consist of:

a) Answers to questions contained in the laboratory outlines unless such questions are obviously answered in the drawings.

b) Additional explanations, original observations, or comments upon any of the laboratory work as the student sees fit.

c) Complete accounts of experiments performed in the laboratory, whether individual or demonstration. This account should follow some such plan as this: (1) purpose of the experiment, (2) observations and results, (3) conclusions and explanations.

4. It is preferable but not compulsory that the notes be written in ink. Write on one side of the paper only.

Additional Miscellaneous Directions

1. The letter (*A*) signifies consult the assistant; (*R*), consult a textbook; (*L*), the matter will probably be discussed in the lectures.

2. Right and left refer to the animal, not to the student, unless so stated.

3. All work is to be individual unless the supply of material is insufficient.

4. When handling small living animals with a dropper be sure that the dropper has never been used for chemicals. Have one dropper plainly marked which you use only for water or other harmless solutions. It has often happened that

the entire supply of animals for the class has been killed because some careless student used a dropper which he had previously employed for chemicals.

In dropping a cover glass upon a slide, first put one edge of the cover glass against the slide until the liquid on the slide comes in contact with the cover glass, then lower the cover glass slowly. In this way, air bubbles under the cover glass are avoided.

Mount living animals in water. The water should be sufficient in amount to reach to the edges of the cover glass but not enough to cause the cover glass to float about. Absorb extra water with a piece of filter paper. While examining living animals be careful that the water does not dry up. When this happens the animals will begin to slow down their movements, to flatten out, and finally to burst. Replenish the water from time to time by placing a drop in contact with the edge of the cover glass.

Living tissues of an animal must be mounted in fluid of the same osmotic pressure as the fluids of the animal. In the case of the frog, this is a .6 per cent solution of common salt.

When dissecting, have the animal firmly fastened to the wax bottom of the dissecting tray with pins. Insert the pins obliquely, not vertically, so that they will not get in your way. Dissection is usually to be carried out under water. Dissect with blunt instruments, forceps, or probe, and have an instrument in each hand, one to hold the part, the other to dissect with. Never cut anything unless you are sure what you are cutting. Dissect lengthwise along blood vessels, nerves, tubes, etc., not across them.

In staining tissues, use a small amount of the stain. If the stain is applied before the cover glass is put on, drop a small drop of the stain on the tissue, then cover. If the stain is applied after the cover glass has been put on, place a drop of the stain in contact with one edge of the cover glass, apply a piece of filter paper to the opposite edge, and draw the stain under by suction. It generally requires a few minutes for a stain to act, especially when living animals are to be stained. If too much stain is used the whole structure becomes uniformly stained, so that the details are blotted out. In this case, make a new preparation or remove the excess stain by running a little dilute acid under the cover glass.

The Use of the Microscope

1. Parts of the microscope. Remove the instrument from the case. It consists of a horse shoe-shaped *base* from which arises a vertical *pillar*, from which extends an arm supporting a vertical, hollow *tube;* below this a square or round *stage*, with a central opening; under the stage, a *condenser;* and below the condenser, a movable *mirror*. At the lower end of the tube is a swinging *nosepiece*, into which the *lenses* are to be screwed. The condenser is a system of

lenses for focusing light upon the stage. On the lower side of the condenser is an *iris diaphragm*, made up of movable leaves; the size of the opening between the leaves can be changed by means of a button located at the side of the diaphragm. Turn the microscope upside down so that you can see the diaphragm and observe the effect of sliding the button back and forth. The mirror has two surfaces, plane and concave. When the condenser is in use, the plane surface of the mirror is to be employed, as the concave mirror converges the light, and thus conflicts with the action of the condenser.

2. The lenses. Find on a shelf in the microscope box, two cylindrical metal cases. Unscrew these and remove from them the *objectives*. These objectives are marked 3 and 7 (German system) or 16 mm. and 4 mm. (American system). Note that 3, or 16 mm., is shorter, has a larger lens, and is therefore of a lower power of magnification than 7, or 4 mm. Screw the objectives into place in the nosepiece, holding each with both hands while doing so to avoid possibility of dropping.

On another shelf in the miscroscope box will be found two *eyepieces* or *oculars*. These are Nos 1 and 3, or 4× and 8×; the first-named one is in each case the lower power. Place the No. 1 or 4× eyepiece in the top of the tube.

3. The adjustment screws. On the arm between the pillar and the tube is a pair of vertical screws. These move the tube considerable distances at each turn, and are therefore designated as the *coarse adjustment*. At the top of the pillar is a horizontal screw, the fine adjustment, which moves the tube only a very short distance at each turn. Turn each of the screws and note the effect on the tube. When the fine adjustment is turned clockwise, the tube moves down; when counter-clockwise, it moves up.

4. To use the low power:

a) Swing the low-power objective into place, and place the low-power eyepiece in the top of the tube.

b) While looking through the eyepiece, turn the mirror toward the light until the field of the microscope becomes suddenly bright.

c) Place the object to be examined, mounted on a glass slide, in the center of the opening on the stage.

d) Lower the tube by means of the coarse adjustment until the objective is close to the slide.

e) While looking into the eyepiece, slowly raise the tube by means of the coarse adjustment until the object comes into view. The object is now said to be in *focus*. Note the distance from the object at which the lower power comes to focus.

f) Adjust the light to the best advantage by means of the iris diaphragm. *This is a very important point.* Students, as a rule, use too much light, which drowns out the details of the object and is hard upon the eyes. Always adjust the light for every object looked at.

g) If the object is very transparent, reduce the amount of light before beginning to focus, because too much light will make it invisible. Moving the slide slightly while trying to focus will facilitate the process.

5. To use the high power:

a) The object must always be under a cover glass when the high power is to be used.

b) The high power cannot be used with thick or thickly mounted objects.

c) The object to be viewed must always be found first with the low power. *Never try to examine anything first with the high power.* Place the object or part of the object which you desire to study with the high power in the exact center of the low-power field.

d) Swing the nosepiece around so that the high-power objective comes into place.

e) The object should now be nearly in focus and is brought into exact focus by means of the fine adjustment. If the object is not in focus when the high power is swung into place, the microscope is not perfectly adjusted and the following procedure must be followed:

(1) The center of the field of the high power may not be the same as the center of the field of the low power. The object will therefore not be in the field at all when the high power is swung around. The remedy is to find out where the center of the high-power field is on the low-power field and to place the object there instead of in the center when preparing to use the high power.

(2) The focus of the high power may not be the same as the focus of the low power. This is the common difficulty. The remedy is to screw the tube up or down after the high power is in place until a focus is obtained. You will have to find out by trial whether to screw up or down. If up, then always remember to raise the tube of the microscope before swinging the high power into place, as otherwise the objective will strike against the slide.

(3) Have the assistant help you find out the peculiarities of your microscope.

f) After getting a focus, adjust the amount of light by means of the diaphragm. The amount of light best for the high power is never that best for the low power.

g) Note how close to the slide the high-power focuses. For this reason thick objects cannot be viewed under the high power, and care must always be taken not to run the objective into the slide, as this will break the slide and may injure the lens.

6. Plane of focus. As the microscope is an optical instrument, the planes of focus of its lenses are geometrical planes, i.e., planes without thickness. All objects viewed through the microscope have an appreciable thickness. It is therefore obvious that *no object, no matter how thin it is,* can be seen in its totality in a *single plane of focus,* as some parts are certain to lie outside that plane. *It is therefore necessary to change the focus continually while viewing an object in the microscope.* This is particularly essential when using high powers. *Students*

almost invariably make the grave mistake of getting an object in focus, and then examining it without any further change of focus, with the result that they do not see all parts of the object, nor get a true idea of the relation in space of the parts. The practiced microscopist never takes his hand from the fine adjustment screw but continually changes the focus as he looks. The student should at once form a habit of doing likewise.

7. Moving the slide. As the image in the microscope is reversed, the slide must be moved in the opposite direction from that in which it is desired to move the image. This will soon become a habit. In moving a slide, *do not put both hands upon it* but grasp it by the edges between the thumb and index finger of one hand. This leaves the other hand free to shift the focusing screw.

8. Miscellaneous directions:

a) If the image is dim or indistinct, or if the field rolls, or if the high power cannot be focused, then in all probability the lens is dirty or wet. Clean it with an old, soft handkerchief. If after cleaning the lens the high power will not focus, then the material under examination is too thick and must be made thinner. In case of any trouble with the microscope, don't tamper with the instrument, but call the assistant.

b) If images of buildings, etc., appear in the field, they may be obliterated by using the concave surface of the mirror, or by lowering the condenser.

c) In working with artificial light, use the concave surface of the mirror.

d) Round black rings in the field are air bubbles under the cover glass.

e) If the fine adjustment turns without producing any effect upon the tube, it has come to the upper limit of its range and must be screwed down. If it will not turn at all, it has come to the lower limit of its range and must be screwed up.

f) Keep the microscope clean and free from dust. Do not let it stand in the sunlight. Do not use sunlight for illumination in looking through the microscope.

g) Keep both eyes open when looking through the microscope. If you find this difficult, try placing an oblong of stiff paper around the top of the microscope, so that the unused eye will not see objects.

9. The magnification of the low power is about 50; of the high power, 500.

KEEP THE LENSES CLEAN

ADJUST THE LIGHT FOR EVERY OBJECT

ADJUST THE FOCUS CONTINUALLY WHILE YOU LOOK

TABLE OF CONTENTS

		PAGE
I. GENERAL SURVEY OF THE FROG		1
A. Killing the Frog		1
B. External Anatomy of the Frog		1
C. The Buccal or Mouth Cavity		4
D. Body Wall, Coelome, Mesenteries		5
E. General Internal Structure		8
II. GENERAL PHYSIOLOGY OF THE FROG		14
A. Function of the Nervous System; Irritability, Conductivity		14
B. Function of the Muscular System; Contractility		15
C. Function of the Digestive System; Digestion, Absorption		16
D. Function of the Respiratory System; Respiration, Oxidation		18
E. Function of the Excretory System; Excretion		19
F. Function of the Circulatory System; Circulation of the Blood		20
G. Function of the Reproductive System; Reproduction, Development, the Life-Cycle		20
H. Summary of Physiological Processes		23
III. GENERAL HISTOLOGY: CELLS AND TISSUES		24
A. Study of a Typical Cell		25
B. Studies of Tissues		26
IV. GENERAL HISTOLOGY: STRUCTURE OF ORGANS		33
A. Structure of the Liver		33
B. Structure of the Intestine		33
C. Structure of the Stomach		35
D. Structure of the Skin		36
E. Structure of the Kidney		37
F. Structure of the Spinal Cord		38
V. THE SPECIAL ANATOMY OF THE FROG		39
A. The Digestive System		39
B. The Urinogenital System		40
C. The Respiratory System		41
D. The Circulatory System: the Venous System		41
E. The Circulatory System: the Arterial System		44
F. The Circulatory System: the Structure of the Heart		45
G. The Nervous System		46
H. The Skeleton		51
I. The Muscular System		55
J. General Anatomical Principles		61
VI. THE PROCESS OF CELL DIVISION		62
A. Mitosis in the Eggs of *Ascaris*		62
B. Mitosis in Plant Root Tips		64
C. Mitosis in the Eggs of the Whitefish		64

INSTRUCTIONS FOR THE LABORATORY ASSISTANTS

INDEX

I. GENERAL SURVEY OF THE FROG

It is believed that the best introduction to the science of zoölogy is the thorough study of a single, relatively complex animal. For this reason, we shall first examine in detail the anatomy, the physiology, and the microscopic structure of one of the commonest animals, the frog. After learning how such an animal is constructed and how it employs the structures which it possesses, we shall be in a position to understand the make-up of other animals, and to appreciate by what changes and alterations these have been built up from an extremely simple starting-point.

A. KILLING THE FROG

The frog may be killed either by placing it for 15 to 20 minutes in a closed vessel with a wad of cotton soaked in ether or chloroform, or by the method of *pithing*. In pithing, a blunt instrument is thrust through the space between the posterior end of the skull and the beginning of the vertebral column and the nervous system is destroyed. To pith a frog, grasp the animal firmly in the left hand and bend the head down by placing the forefinger across the snout. With the finger or a blunt instrument feel for the depression between the posterior end of the skull and the first vertebra (it is located at about the level of the fore limbs). Cut through the skin at this place with a scissors, and press firmly upon the depression with a blunt instrument until the instrument breaks through the muscles into the cavity of the skull. Then thrust the instrument forward into the brain and then backward into the spinal cord, moving it about so as to mash the nervous system thoroughly. The best instrument for pithing is a stout wire hairpin or a blunted hatpin. If the pithing is properly done, spontaneous movements cease.

After either pithing or etherization, it will be noted that many of the activities of the frog continue; the heart keeps on beating for a considerable length of time and movements of various kinds can be elicited by the proper procedure. Is the frog dead? What is meant by death in the higher animals? Do all parts of an animal die at the same time? What part of the frog is really dead? (*A.*)

B. EXTERNAL ANATOMY OF THE FROG

Obtain an etherized frog, place it in the dissecting pan, and carefully note the following points. Read also Holmes, chapter iii. The *body* of the animal consists of a flattened *head* and a short somewhat spindle-shaped *trunk*. There is neither *neck* nor *tail*. Head and trunk (also neck and tail when present) constitute the *axial* part of an animal while the limbs are designated as the *appendicular* part. The head is the *anterior* end; the opposite region of the body is the *posterior* end; and the terms "anterior" and "posterior" are also employed to indicate the relative positions of other structures with reference to head or tail, as, for instance, one would say that the fore limbs are anterior to the hind limbs. The back or upper surface is the *dorsal* side; the lower or under surface, the *ventral* side; and the regions between these are referred to as *lateral*. The middle of the dorsal side is the *median dorsal line*, and that of the ventral side, the *median ventral line*. A plane passing through these two lines from anterior to posterior end is the *median sagittal plane*, and divides the animal into *right* and *left halves*. If these two halves are identical or nearly so, the animal is said to be *bilaterally symmetrical*. Is the frog bilaterally symmetrical as far as you can observe from its external anatomy? Is man? There are several other types of symmetry among animals, one of which will be met with later in the course. An imaginary axis in the sagittal plane is the *antero-posterior axis* or *sagittal axis*. Any axis in the median plane from the dorsal to the ventral side is a *dorsoventral axis;* from the median plane to the sides, a *medio-lateral axis*, etc.

These terms are applicable to the vast majority of animals, and the structures of animals are arranged with reference to such *planes* and *axes of symmetry*.

Note differences in color between the dorsal and ventral surfaces, and observe whether the *color pattern* is the same on all individuals. Read Holmes (p. 37), on the power of frogs to alter their color.

The head ends in a triangular *snout* which incloses the relatively large *mouth cavity*, and which bears on its anterior extremity two small openings, the *nostrils* or *external nares*. These, unlike our own, can be opened and closed by lowering and elevating certain bones of the upper jaw (Holmes, p. 171). Posterior to the nares are the large prominent *eyes*, in which may be distinguished the golden *iris*, surrounding a central opening, the *pupil*. The eye is provided with the following eyelids, as should be determined by lifting each with a forceps: an *upper eyelid*, a well-developed fold of skin which covers the upper portion of the eyeball; a *lower eyelid*, semicircular in shape, representing scarcely more than the free edge of the skin; and the *nictitating membrane*, a thin, transparent, very extensible membrane, which is really an outgrowth of the lower lid. A vestige of the nictitating membrane is present in our own eyes as a small crescent-shaped fold near the inner corners.

Obtain a living frog, gently touch an eyeball and observe the action of the eyelids. Which eyelids are movable? Stimulate the eyeball more strongly and observe that the whole eye can be dropped down into the mouth cavity. The socket in the skull which holds the eye is designated as the *orbit*.

Returning now to the dead specimen, note a circular area of tense skin just posterior to the eye. This is the *tympanic membrane* or *drum membrane* of the ear, which covers the cavity of the *middle ear*. The *external* ear and the passage leading in from it, prominent in ourselves, are entirely wanting in the frog. Near the center of the drum membrane, a small projection may usually be noticed; this is the end of the *columella*, a small bone which transmits inwardly the vibrations of the tympanic membrane.

In the median line of the head, slightly anterior to the level of the eyes, a small light-colored spot, the *brow spot*, may usually be found. In dark individuals, however, it may be concealed by pigment. In the embryonic development of the frog, this spot is in connection with a portion of the brain, called the *pineal body*, and it itself is the useless vestige of a former third medially located eye (Holmes, p. 64).

On the dorsal side of the trunk extending posteriorly from the eyes note two light-colored ridges, where the skin is much thickened owing to the presence of large poison glands underneath. These ridges are called the *dorsolateral dermal plicae*. At the posterior end of the trunk on the dorsal side between the bases of the hind legs is a small opening, the *anus*, which is the end of the digestive tract. In the middle of the back a characteristic hump is present, owing to an alteration at this place in the structure of the vertebral column (see a dried skeleton, or consult Holmes, Fig. 63, p. 230).

The *fore limb* is short and consists of three divisions, *upper arm*, *forearm*, and *manus* or *hand*. How many fingers has the hand? To which of your fingers do these correspond (Holmes, p. 65)? The rudiment of the missing finger may be felt under the skin on the inner side of the hand at the base of the present first finger, and may be seen on the skeleton of the hand (*A*). In the male frog, the inner finger is enlarged and swollen at the base, especially during the breeding season.

The *hind limb* likewise consists of three parts, *thigh*, *shank*, and *pes*, or *foot*. The *ankle* is remarkably elongated. There are five toes and a rudiment of sixth, called the *prehallux*, may be felt on the inner side of the smallest toe, which corresponds to our great toe. Such additional fingers occur not infrequently among the vertebrates, but their morphological significance is unknown.

The skin is smooth, slimy in life, and possesses neither hairs, scales, nor claws. Note that in general it is more loosely attached to the body than in most animals.

Make an accurate drawing of the frog from the dorsal side, putting in all of the structures and parts which have been mentioned. Before beginning to draw re-read carefully the directions about drawings. Label this and all subsequent drawings fully.

NOTE.—At the close of the first laboratory period, make an incision about one-half inch long through the skin only of the left abdominal wall of your frog and place the animal in the jar of preserving fluid which will be assigned to you. The animal *must* always be kept in this jar when not in use. It *must never* be left out on the tables, and never allowed to become dry.

C. THE BUCCAL OR MOUTH CAVITY

Note that the two jaws fit together very tightly; this is essential for respiration, as will be seen later. Open the mouth to its widest extent, cutting the angle of the jaws if necessary (A). Consult Holmes, chapter iv, pp. 68–73.

1. **Roof of the mouth cavity.**—The edge of the upper jaw is covered by a fold of skin, the *upper lip fold.* Just within this fold and concealed by it is a semicircular row of teeth, the *maxillary* teeth, borne upon the edges of the *maxillary* and *premaxillary* bones of the skull. It is necessary to grasp the lip fold with a forceps and pull it outward in order to reveal the teeth. Obtain a dried skull, examine the teeth with a hand lens, and note the following points. The teeth are not set into *sockets*, that is, cavities in the jaw, as are our own teeth, but they are fused by their bases and sides to the margins of the jaw, only their tips projecting freely above the edge of the bone. Each tooth is a hollow cone, consisting of two parts separated by an indistinct groove: an upper part, the *crown*, composed of *dentine*, and coated externally with a shiny material, the *enamel;* and a lower part, the *root*, composed of *cement.* The teeth are replaced when lost and two or three sets of new teeth may usually be seen at the bases of the old ones.

Returning now to your own specimen, locate within the row of maxillary teeth a deep groove, the *sulcus marginalis.* At the tip of the jaw this groove deepens into a pit, the *median subrostral fossa;* on each side of this is an elevation, the *subrostral pulvinar* or *cushion;* and lateral and adjacent to this is another depression, the *lateral subrostral fossa.*

Within the *sulcus marginalis* is the *roof* of the mouth cavity, properly speaking. The anterior extremity of this is occupied by two oval openings, the *internal nares* or *choanae;* they are the internal openings of the cavities of the nose, whose external openings were already noted. The two pairs of nares, therefore, with the cavities of the nasal chambers, constitute the *respiratory passage* through which air is drawn into the *buccal cavity.* Between the choanae are two patches of *vomerine teeth,* located upon the *vomer* bone. The greater part of the roof of the mouth is occupied by two large rounded prominences, where the eyes are located, and, as already observed, the eyes can be withdrawn into the mouth cavity. At each side of the posterior end of the roof is an opening, the entrance to the *Eustachian tube* (*auditory tube* in more recent terminology). Where does it lead? Consult Holmes, p. 69.

2. **Floor of the mouth cavity.**—The edge of the lower jaw forms a ridge which fits into the sulcus marginalis. At the tip of the lower jaw is an elevation, the *prelingual tubercle*, and on each side of this a depression, the *prelingual fossa.* Note how exactly these fit into the elevations and depressions of the upper jaw. Are teeth present on the lower jaw? The greater part of the floor of the mouth is occupied by the *tongue.* Note the size, shape, and attachment of this organ, and find out how it is used in catching prey (Holmes, pp. 26, 70–71). Turn the

tongue forward and feel the floor of the mouth behind the tongue. It is stiffened by a cartilaginous plate, the *body* of the *hyoid*, whose anterior end is hollowed out to receive the base of the tongue, and gives off a pair of slender curving processes, the *anterior horns* of the *hyoid*, which extend posteriorly to the ears. Scrape away the membrane from the floor of the mouth in order to see these structures more clearly. Posterior to the body of the hyoid in the median line is a circular hardened elevation, the *laryngeal prominence*, which bears in its center an elongated slit, the *glottis*. Where does the glottis lead? At the back of the mouth cavity, roof and floor converge to a large opening, the beginning of the *esophagus*, the first portion of the digestive tract. In the male frog, the slitlike opening of the *vocal sac* is present on each side of the floor near the edge of the jaw, on a level with the glottis. On the use of the vocal sacs in croaking, see Holmes (p. 167).

Draw the floor and roof of the mouth.

D. BODY WALL, COELOME, MESENTERIES

1. **Structure of the body wall** (see Holmes, chap. iv, pp. 73–80).—Remove the skin slowly from the trunk of the frog, noting carefully at what places the skin is attached to the underlying parts by means of weblike partitions. (While doing this note also the blood vessels to the skin described in the next paragraph, and in the median dorsal line, the *sensory nerves* passing in pairs from the skin into the vertebral column.) The space under the skin is divided by these partitions into compartments, called the *subcutaneous lymph spaces*, or *lymph sacs*, which in life are filled with a fluid, the *lymph*. Compare your observations with Holmes (Fig. 78, p. 281), and read what Holmes says about the lymphatic system. The lymph is similar in composition to and derived from the blood, except that it contains no red blood corpuscles. It is in direct contact with the living substance to which it supplies food and oxygen, and from which it removes waste products. The frog and its relatives differ from other vertebrates in this enormous development of huge lymph spaces, not only under the skin, but also throughout the body. This structural feature is probably associated with the amphibious habits of these animals.

In removing the skin, note the extensive supply of blood vessels to the skin, and particularly the following two large vessels: the *musculo-cutaneous* vein, which runs posteriorly in the muscles of the ventro-lateral region of the body wall, and then turns and passes to the skin in the partition between the *abdominal* and *lateral* lymph sacs; and the *cutaneous* artery, which emerges in front of the shoulder and supplies the skin of the dorsal side. This relatively large development of skin blood vessels is due to the respiratory function of the skin, to be discussed more fully later.

The removal of the skin exposes the *muscles* of the body wall and the *skeleton* which they inclose and to which they are attached. The principal parts of the skeleton may be felt imbedded in the muscles. Identify them as follows with the aid of Holmes (Fig. 63, p. 230). The head contains a bony case, the *skull*, to which the upper jaw is immovably fused, while the lower jaw is hinged to it by a joint. A bony arch, the *pectoral girdle*, supports the fore limbs. Feel this on both dorsal and ventral sides at the level of the fore limbs. On the dorsal side, the girdle terminates in a flat thin bone with a cartilaginous border, called the *suprascapula*. Ventrally, the bones of the girdle, covered by the *pectoral* muscles, are seen to be articulated to a slender chain of bones and cartilages which occupies the median ventral line and ends anteriorly and posteriorly with conspicuous rounded cartilaginous expansions. This whole structure is called collectively the *sternum*, or *breastbone* (see Holmes, Fig. 67, p. 239). The hind limbs are similarly supported by the *pelvic girdle*. Feel for this on both sides; ventrally it forms a hard crest between the bases of the hind legs; dorsally two of its bones, the *ilium* bones, extend forward, producing the two conspicuous lateral ridges on the lower half of the back. The median dorsal line is depressed, forming a groove. Pass the point of an instrument along this groove and feel the *vertebral column* lying underneath, and its lateral expansions, the *transverse processes*. The last pair of these has enlarged bulbous ends, which are really rudimentary *ribs*, to which the anterior ends of the *ilium* bones, mentioned above, are firmly attached. It is this region which produces the external hump in the frog. Posterior to this point, the median line is occupied by a long slender bone, which has been produced by the fusion of a number of vertebrae; it is called the *urostyle*.

The lateral and ventral abdominal walls are supported by *muscles* only, skeleton being absent. There are several layers of these muscles, the external layer consisting of, laterally, the *external oblique* muscle; ventrally on each side of the median line, the segmented *rectus abdominis* muscle extending from the pelvic girdle to the sternum; and between these, and partially covering the external oblique, a posterior slip from the *pectoral* muscles. A flat, thin muscle, the *mylohyoid*, extends transversely across the ventral side of the lower jaw. This arrangement of the ventral musculature is very similar to that of all vertebrates, including man. The median ventral line from the pelvic girdle to the posterior expansion of the sternum is occupied by a white strip, the *linea alba*, under which there runs, in the frog, a conspicuous blood vessel, the *anterior abdominal* vein. For further details and a picture of the foregoing features, see Holmes (Fig. 70, p. 249).

Cut through the muscles on the ventral side *to the left (frog's left) of the linea alba* from a point just in front of the pelvic girdle up to the mylohyoid muscle, cutting through the pectoral girdle. A large cavity, the *body cavity, coelome*, or *pleuroperitoneal cavity*, which contains the internal organs, or *viscera*, is thus exposed. This space is lined by a smooth shining membrane, the *pleuroperi-*

toneum (frequently but less correctly called for brevity *peritoneum*). It should be recalled that in man the body cavity is divided by means of a muscular partition, the *diaphragm*, into two completely separated portions, an anterior *thoracic* cavity and a posterior *abdominal* cavity. The lining membrane of the former is then called *pleura*, and that of the latter, *peritoneum*, in the correct sense. Since, however, in the lower vertebrates, including the frog, a diaphragm has not yet evolved, one continuous coelome, or pleuroperitoneal cavity, is present, and its lining membrane is named the pleuroperitoneum.

The *body wall* thus consists of three layers: the skin, the muscles with their contained skeleton, and the coelomic lining, or peritoneum, with the subcutaneous lymph spaces lying between the first two layers.

2. **The peritoneum and mesenteries.**—The peritoneum not only lines the coelome, but forms a close investment of all the viscera, for which purpose it is frequently pulled away from the body wall as a double-walled membrane. That portion of the peritoneum which adheres to the inside of the body wall is called the *parietal peritoneum;* that which invests the viscera is the *visceral peritoneum,* or *serosa;* and that which extends from the body wall to the individual organs or from one organ to another is a *mesentery* or *ligament.* In the formation of a mesentery, the peritoneum leaves the body wall, passes over the surface of the organ, and returns to the body wall at the same point from which it left, producing a double-walled membrane between the organ and the body wall. It is thus evident that the peritoneum is everywhere continuous and unbroken, and that the viscera are really outside of the peritoneum, which forms a closed sac into which the viscera appear to be pushed from without. The condition is not really brought about in this way but by the fact that the peritoneum develops later than the viscera and closes over them after they have formed. The visceral peritoneum is so tightly applied to the surface of the viscera that it cannot be separated from them.

Extreme caution must be used in examining the following mesenteries, especially those in the region of the heart, so as not to destroy them.

Note the following mesenteries. Lift up the pectoral girdle cautiously and find beneath it a thin-walled sac, the *pericardial sac,* which contains the heart. Pick up the pericardial sac gently with a forceps and observe that it is separated from the heart by a space, the *pericardial cavity,* in which the heart moves freely. The heart is in reality, like the other viscera, inclosed in a double sac; the inner one tightly invests the heart, constituting in fact a serosa, or *visceral pericardium;* the outer sac as already noted is loose and separated from the heart by the pericardial cavity, forming a *parietal pericardium.* The pericardium is therefore a part of the genera lining of the coelome, and the pericardial cavity is a part of the coelomic cavity, from which, however, it has become completely separated during embryonic development, by the formation of the pericardial sac (parietal pericardium).

The general pleuroperitoneal membrane in the region of the heart leaves the body wall in the median dorsal line and passes ventrally on each side of the pericardial sac and in inseparable contact with it; the two sides then meet ventrally below the heart and form a double membrane which extends vertically from the pericardial sac to the ventral median line under the sternum. This vertical membrane also supports the *liver*, the large reddish-brown organ around and posterior to the heart, and this part of it is therefore known as the *suspensory* or *falciform ligament* of the liver. Follow the suspensory ligament posteriorly along the median ventral line and note that it supports the large anterior abdominal vein which was mentioned in the preceding section. Follow along the median ventral line and at the posterior end of the coelome locate the *median ligament* of the *bladder*, a mesentery which attaches the *urinary bladder*, a thin-walled, often shriveled sac, to the ventral body wall.

There was originally in the embryo, a complete double-walled mesentery running from the median dorsal line to the median ventral line, and inclosing the viscera between its two walls. The coelome was thus divided into two entirely separated halves. The portion of this mesentery between the digestive tract and the dorsal wall is called the *dorsal mesentery*, and that from the digestive tract to the ventral wall, the *ventral mesentery*. The dorsal mesentery, as will be seen shortly, is still intact in the adult frog, but the ventral mesentery has entirely disappeared, except for certain remnants already mentioned—the suspensory ligament of the liver, and the ligaments of the bladder, and certain ligaments running from the liver to the intestine, which will be described below.

Cut through the ventral mesenteries, and pin out the body wall, making cross cuts if necessary so that the viscera will be fully exposed.

E.   GENERAL INTERNAL STRUCTURE

The frog is made up of a number of definite structures called *organs*, each of which has a definite function to perform. All of the organs which aid in performing the same function are grouped together as a *system* or *tract*. The organs constituting one system may be all alike or may be different among themselves. In general in a complex animal there are ten systems: *skin* and its derivatives, *skeletal, muscular, digestive, circulatory, respiratory, excretory, reproductive, nervous,* and *sensory systems.* To this list, there should probably be added, in the case of vertebrates, an eleventh, composed of a number of glands, which were originally derived from the other systems, but have lost connection with them and have taken on peculiar but extremely important functions. This group of glands is spoken of collectively as the *glands of internal secretion*, also as *cryptoretic* or *endocrinous organs.* Attention has already been called to the muscular and skeletal systems; the other systems will now be described briefly and will be studied in detail later (Holmes, chap. iv, pp. 73-80).

1. **Circulatory system.**—This consists of the *heart*, the *arteries* (vessels leaving the heart), the *veins* (vessels entering the heart), and the *capillaries* (microscopic vessels between the ends of the arteries and the beginnings of the veins). Remove the pericardium by cutting it off with a fine scissors, and examine the heart. The chambers of the heart are called *ventricle, auricles, sinus venosus,* and *conus arteriosus*. The ventricle is the posterior, thick-walled, conical portion, the point of the cone being designated as the *apex* of the ventricle, and the base of the cone, the *base*. The auricles are the two dark-colored, thin-walled sacs anterior to the ventricle. Extending from the right side of the base of the ventricle obliquely forward between the auricles is a tube, the conus arteriosus, which forks into two trunks leading away from the heart (Holmes and most other textbooks erroneously refer to this chamber of the heart as the *bulbus arteriosus*). To locate the sinus venosus, turn the heart up so that the apex points anteriorly, and, putting the heart on a stretch, identify a small chamber appearing as a dorsal and posterior continuation of the auricles, from which, however, it is separated by a distinct white line. Three veins, dark red tubes, will be seen emerging from the liver to enter the sinus venosus at its posterior border, and each of its sides receives a vein which runs along the margin of the adjacent auricle. Through these large veins all of the venous blood in the body is returned to the sinus venosus which passes it on into the right auricle. Note the membrane by which the pericardial sac is attached to the serosa of the liver; this is the *coronary ligament* of the liver.

2. **Respiratory system.**—This system consists of the *glottis*, noted in the study of the floor of the mouth, a pair of *lungs*, and the *larynx*, a chamber connecting the lungs with the glottis. The lungs will be found attached to the anterior wall of the coelome, lateral to the heart. Push the liver and other structures aside in order to see them. Each is closely invested by a sac of peritoneum. The larynx will be studied later.

3. **Digestive system.**—Its parts are the *esophagus, stomach, small intestine, large intestine,* and *digestive glands*. The esophagus lies dorsal to the heart and will be seen more clearly at a later time. It passes into the elongated cylindrical stomach, a conspicuous white organ on the left side of the body dorsal to the liver. (If the animal is a female, the large *ovaries*, voluminous lobed black and white masses, will obscure the rest of the abdominal viscera, and may be removed, at least on the left side.) From the end of the stomach, trace the small intestine, a coiled tube, to its enlargement into the large intestine. Locate the *urinary bladder*, a thin-walled, usually shriveled sac at the extreme posterior end of the coelome. The large intestine passes through the bony ring formed by the pelvic girdle, and opens to the exterior through the anus. The entire tube from mouth to anus is the *alimentary canal*. Associated with the alimentary canal are two digestive glands, the *liver*, already noted, and the *pancreas*. The latter is a

... very irregular body, lying in the mesentery which extends between ... the small intestine, and the stomach. The first part of the small ... called the *duodenum*, will be found to bend abruptly forward ... the liver, and in the mesentery between this bent portion of the ... intestine and the stomach is located some yellowish branching material ... constitutes the pancreas. Between the lobes of the liver lies the round ... *gall bladder*.

The following mesenteries should be carefully identified, and each organ ... up with a forceps as you read the description, and its relation to the body ... and other organs noted. The entire alimentary canal is suspended from the ... dorsal line of the coelome by an extensive mesentery, the *dorsal mesentery,* in which run the blood vessels, lymphatics, nerves, etc., that supply the digestive ... Pick up the stomach and intestine with a forceps and see how they are ... to the dorsal median line by this mesentery, a very delicate and trans-... membrane. The part of the mesentery which supports the esophagus is called the *meso-esophageum;* the stomach, *mesogastrium,* or *mesogaster;* the small intestine, *mesenterium,* or *mesentery* proper; and the large intestine, *mesorectum.* The small red body located in the mesorectum is the *spleen,* an organ associated with the lymphatic system. Pick up the urinary bladder with a forceps, and find its attachments to the ventral median line by the *median ligament* of the bladder, already noted; to the large intestine by the *rectovesical ligament;* and to the lateral body wall on each side by the *lateral* ligaments of the bladder. The liver is connected to the dorsal wall of the coelome by the *dorsal mesentery* of the liver or *mesohepar;* to the pericardial sac by the *coronary* ligament; to the digestive tract mainly by the *hepato-gastro-duodenal* ligament in which the pancreas is located; and to the ventral wall in the median line by the *suspensory* ligament, previously noticed and cut. The student will perceive that the ligaments of the bladder, and the suspensory and hepato-gastro-duodenal ligaments are remnants of an originally more extensive *ventral mesentery,* which connected the alimentary canal and all structures ventral to it to the ventral body wall.

The mesenteries constitute an ingenious device which permits the viscera to adjust themselves to movements of the body and to carry out their own movements freely, and yet at the same time holds them in place with reference to each other and to the body wall.

4. Reproductive system. This system consists of a pair of reproductive glands, or *gonads,* in which the sexual elements are produced, and a pair of *ducts* which convey these elements to the exterior.

The female gonads, or *ovaries,* are, in the mature female, the most conspicuous organs in the body. Each is a large, lobed mass composed mainly of numerous black and white eggs, or *ova.* Each is suspended from the dorsal wall by a mesentery, the *mesovarium*. Lateral to each ovary is its duct, the *oviduct*, a conspicuous white, much-coiled tube, also supported by a mesentery, the *mesotubarium*.

The male gonads, or *testes*, are a pair of small, oval yellow bodies situated close to the dorsal body wall near the median line. The intestine must be pushed away to see them. Each has a short mesentery, the *mesorchium*. The testes have no ducts but the male reproductive elements pass to the exterior through the ducts of the kidneys, which will be identified in the next section. In our common species of frog, *Rana pipiens*, the male possesses a vestigial oviduct, a distinct though small white tube running along the lateral border of each testis.

Attached to the anterior end of each testis, and in a similar position in the female frog, is a *fat body*, consisting of a tuft of yellow, finger-shaped processes. This organ is a storehouse for nutritive material, and its size varies with the physiological condition of the frog, being very small in the spring and very large in the fall before hibernation begins.

5. **Excretory system.**—It is composed of a pair of *kidneys* and their ducts. Turn all of the abdominal viscera to the right, and follow the peritoneum along the left lateral wall of the coelome around to the dorsal side. Note that the peritoneum leaves the body wall dorsally and stretches as a thin membrane across the dorsal side of the body cavity, leaving a large space between itself and the muscles of the dorsal wall. This space is called the *subvertebral lymph sinus*, or *cisterna magna*, and, like the subcutaneous lymph spaces, it is a part of the extensive lymphatic system which the frog possesses. Within the cisterna magna, certain structures, including the kidneys, are located. Such structures are said to be *retroperitoneal*, i.e., they lie behind the peritoneum. Break through the peritoneum which forms the ventral wall of the cisterna magna, and locate in the cavity of the cisterna magna the kidneys, a pair of elongated, flat, red bodies situated close to the peritoneum, which passes across their ventral faces. From the lateral posterior edge of each kidney arises its duct, the *ureter*, or *Wolffian duct*, which empties into the large intestine. The ureters not only carry the urine from the kidneys to the exterior, but also transport the male reproductive elements. Owing to this close connection which exists between the excretory and reproductive systems in all vertebrates, the two systems are commonly referred to as the *urinogenital* system. The urinary bladder, although functionally a part of the excretory system, is morphologically a saclike outgrowth of the ventral wall of the large intestine.

6. **Glands of internal secretion.**—Under this head are gathered together a number of glandlike bodies which secrete into the blood certain substances of very great physiological importance. The *adrenal* gland forms a bright yellow stripe on the ventral face of each kidney. The spleen, which *may* be a gland of

internal secretion as well as a lymphatic organ, has already been noted. The *pseudothyroids* are a pair of small round reddish masses, one on each side a little anterior to the heart, at about the level of the posterior end of the body of the hyoid. A *thyroid* gland is located directly under each pseudothyroid, much deeper down in contact with the hyoid cartilage (see Holmes, Fig. 60, p. 222). It is generally difficult to identify the thyroid glands with certainty. For the location of other ductless glands of the frog and for a discussion of their functions consult Holmes, chapter xii, p. 219.

7. **Nervous system.**—The nervous system is divisible into three parts: the *central nervous system,* consisting of the brain and the spinal cord; the *peripheral nervous system,* consisting of the nerves which pass from the brain (*cranial* nerves) and from the spinal cord (*spinal* nerves) to all parts of the body; and *the sympathetic system,* an outgrowth of the central nervous system, differentiated to control the involuntary activities of the body (as digestive tract, heart, etc.).

Turn the frog back upward and remove the muscles from the head and median portion of the back. This exposes the skull in the head, and a row of spinelike projections in the median line of the back, which are the *neural arches* of the vertebrae. Between the posterior end of the skull and the first vertebra is a space. Insert one blade of a fine scissors in this space, keeping the point well up against the bone to avoid punching it into the soft brain underneath, and cut away the roof of the skull. The white-lobed *brain* lying in a cavity in the skull is thus revealed. Its parts will be studied later. Similarly, cut posteriorly through the arches of the vertebrae, first on one side, then on the other, removing their median portions, piece by piece. The neural arches of the vertebrae are thus seen to inclose a continuous space, the *neural canal,* in which the *spinal cord* is situated.

The cranial nerves cannot be observed at present. The spinal nerves arise in pairs from the spinal cord at regular intervals, one pair emerging between two successive vertebrae. Turn the frog ventral side up and look into the cisterna magna. The stout white cords which appear here closely applied to the muscles of the dorsal body wall are the most posterior spinal nerves.

The sympathetic nervous system consists mainly of a nervous strand on either side of and ventral to the spinal column. At regular intervals along these strands there occur enlargements, or *ganglia,* each of which connects by means of a nerve, the *ramus communicans,* with the adjacent spinal nerve. Have the assistant help you find the sympathetic strands and ganglia in the roof of the cisterna magna, alongside the dorsal aorta.

8. **Sense organs.**—The principal sense organs, the *olfactory sacs,* the *eyes,* and the *ears,* have already been noted. In addition there are sense organs in the skin, which are sensitive to touch, light, chemicals, and differences of temperature, and taste organs in the lining membrane of the mouth.

Draw a diagrammatic cross-section through the frog at the level of the stomach. This diagram must include the skin, the subcutaneous lymph spaces, the vertebral column and contained spinal cord, the cisterna magna and its contents, the stomach, duodenum, pancreas, tips of the lobes of the liver, reproductive organs and their ducts, *and the relation of the peritoneum and the mesenteries to these organs.* This drawing is to be constructed from the knowledge gained from the foregoing examination of the frog. Holmes, Fig. 12, may be consulted but is not to be copied.

The general anatomical features and relations expressed in such a diagram are those common to all vertebrates (except the subcutaneous lymph sacs). The general systems of organs are common to all animals except the lowest, although the details vary considerably.

II. GENERAL PHYSIOLOGY OF THE FROG

In the preceding section a general survey of the systems of organs which make up the structure or *morphology* of an animal has been made. We may next logically consider the functions or *physiology* of these organs, using, as before, the frog merely as an example to illustrate phenomena which are common to all animals.

A. FUNCTION OF THE NERVOUS SYSTEM; IRRITABILITY, CONDUCTIVITY

Irritability is the capacity of living matter to undergo a change as a consequence of changes external or internal to itself. The change in the living organism is known as the *reaction* or *response*, and in the case of animals the response generally becomes visible as a movement. The change which produces the response is the *stimulus;* the act of applying a stimulus to an organism is *stimulation*. When the response appears at a different point than that to which the stimulus was applied, it is quite obvious that conduction of the stimulus has occurred, and this capacity of living substance to transmit stimuli is known as *conductivity*. The time which elapses between the application of the stimulus and the visible response is called the *reaction time*, and is evidently dependent upon conductivity. The nervous system is the irritable and conductile system *par excellence* of the body.

Obtain from the assistant a frog pithed in the brain only (why?). Suspend it by a wire through the lower jaw from the crosspieces of the electric lights. Wait until it hangs quietly. Have a pan or dish of tap water handy to keep the frog moist and to wash off the acid in the following experiment. *Do not allow the frog to become dry.* Dip a *very small* piece of filter paper, not more than 2 mm. square, into dilute acetic acid and stick it to the skin of the abdomen of the frog. Response? Determine with a watch the approximate reaction time. Wash off the acid with water and repeat in various ways, putting the acid on the toes, skin of the hind legs, back, etc. Which part of the body is the most sensitive, as determined by the reaction time? When are the fore legs used? Does the reaction appear to be purposive? Under the conditions of the experiment can it be so? Read Holmes, pp. 300–2. Such a reaction is called a *reflex*, and this particular one is known as the "wiping reflex." The complete path involved in such a reflex can be understood only after a more detailed study of the nervous system. The steps involved are: stimulation of the sense organs in the skin by the acid; conduction of the stimulation along the spinal nerves leading from these sense organs to the spinal cord; conduction in the spinal cord to a level where the nerves to the muscles of the hind legs originate; an impulse from this level of the cord along the nerves in question to the particular muscles needed; contraction of these muscles producing movements of the hind legs.

In this and all subsequent experiments, make careful observations and take notes on what happens and write up the experiment later in your notebook, according to the plan suggested in the introduction.

B. FUNCTION OF THE MUSCULAR SYSTEM; CONTRACTILITY

Contractility is the capacity of living matter to shorten itself. It is probable that all kinds of movements in animals are due to this property, which is particularly specialized in the muscles.

1. **Contractility of voluntary muscle** (muscle under control of the will).— Remove the hook from the frog's jaw and now pith the spinal cord (*A*). Lay the frog back upward, make a circular incision through the skin completely around the base of the thigh, and grasping the cut edge of the skin, completely strip the skin from the hind leg. On the dorsal side of the thigh, three muscles will be seen: laterally, the *glutaeus magnus;* medially, the *semimembranosus;* and between them the small slender *ileo-fibularis.* On the back of the shank is a large spindle-shaped muscle, the *gastrocnemius.* Carefully separate the ileo-fibularis from the semimembranosus and locate between them the *sciatic* nerve, appearing as a stout, white cord running alongside of a dark-colored blood vessel. Carefully isolate the nerve from the blood vessel, handling it with the utmost care. Lift it up and while watching the gastrocnemius muscle cut through the nerve. What happens? What is the stimulus? How does the stimulus get to the muscle? How does the reaction time compare with that in the preceding experiment? Why? The experiment may be repeated as many times as desired by again cutting the nerve between the first cut and the muscle. Satisfy yourself that in the motion the muscle actually becomes shorter and thicker, and that this change in its shape is the cause of the movement.

2. **Contractility of involuntary muscle** (muscle not under control of the will, found mainly in the walls of the digestive tract).—Turn the frog ventral side upward and cut through the ventral body wall to the left of the median line from the pelvic girdle up through the pectoral girdle. Gently pull the stomach out so that it will be clearly exposed and with a forceps pinch the wall of the stomach. Wait for the response. How does the reaction time compare with that of voluntary muscle ? Watch the contraction travel. In which direction does it go ? This kind of contraction is called *peristalsis.* What is its purpose?

3. **Contractility of heart muscle.**—Free the heart carefully from the pericardial sac and observe that the beating of the heart is nothing but a rhythmical contraction of its muscular walls. In what order do the parts of the heart beat ? Observe changes in color, form, and size of the auricles and ventricle during

contraction. Turn the heart with the apex forward so that the sinus venosus can be seen. Does the sinus beat? What time relation does its beat bear to that of the auricles and ventricle?

Count and record the number of heart beats per minute. Then cool the heart by placing small pieces of ice around it, and after it has become thoroughly chilled again count the rate of the heart beats. This illustrates in a striking way the effect of change of temperature on the activities of living matter.

4. Ciliary motion.—Open the mouth of the frog, swab it out with water, and if necessary cut the angle of the jaws to keep it open. Lay the frog ventral side up and sprinkle a little powdered carmine or place small bits of cork on the posterior part of the roof of the mouth. Observe that the particles travel as if carried on a current. (If the experiment fails it is because the frog has been pithed too long, and a fresh frog may be necessary.) The current is due to a multitude of microscopic hairlike processes, called *cilia*, which cover the roof of the mouth, and by their co-ordinated beating produce a current of mucus which is sufficiently strong to carry fairly large particles. Prove that the particles will be transported against the force of gravity by repeating the experiment with the head of the frog tilted at an angle, so that the particles must be carried uphill. In which direction does the ciliary current run and what is its purpose? While the causes of ciliary motion are not at all understood, it is probable that like muscular movement it is a form of contractility.

5. Amoeboid movement.—In this type of movement, which is limited to very small or microscopic masses of living matter, locomotion is accomplished by the flowing out of portions of living substance into processes. The remainder of the substance then flows into the processes, new processes are put out, and a slow change of position is thus effected. Amoeboid movement is probably the most simple form of contractility, but an adequate analysis of its causes has not yet been made.

Amoeboid movement is illustrated in the frog in the black *chromatophores* or color bodies of the skin, and in the *white blood corpuscles*, both of which objects will be seen later. For the present see Holmes (Fig. 49, p. 189, and Fig. 71, p. 259), as neither of them is very favorable for the study of this kind of movement. Amoeboid movement will be studied in the amoeba, the animal from which the movement takes its name.

C. FUNCTION OF THE DIGESTIVE SYSTEM; DIGESTION, ABSORPTION

The food of animals in general consists of water, salts, and organic substances, these latter being divided into three classes, *proteins, carbohydrates*, and *fats*. Examples of proteins are meat, white of egg, curdle of milk, blood clot; sugars and starches are carbohydrates; butter, fat of meat, cream, are examples of fats. While the water and salts pass into the substance of the animal without alteration, the carbohydrates, fats, and proteins are too complex to be absorbed and used by the animal, and must be broken down into simpler substances before they can be utilized. The splitting of these organic foods into simpler, utilizable substances is the process of *digestion*, and the performance of this process is the function of the alimentary tract and its glands. Digestion is brought about by means of certain substances, called *enzymes*, which are manufactured in the walls of the alimentary tract and in the digestive glands, chiefly the pancreas. By means of these enzymes, organisms are able to produce chemical changes in foods which cannot be imitated in the laboratory at all or which can be imitated only by the use of boiling temperatures and reagents which would be fatal to life. Neither the chemical nature of enzymes nor the mode of their action is known, but it is probable that they attach themselves either physically or chemically to the molecules of the substance upon which they act, thus upsetting the equilibrium within the molecule, and causing it to fall into fragments, whereupon the enzyme is set free again unchanged. In general each enzyme is capable of acting upon only one substance or class of substances. Thus, enzymes which split up proteins are called *proteases;* those which split starches and sugars are *diastases;* and those which split fats are *lipases* (see Holmes, chap. vii, pp. 134–38, 142, 156, 163).

In the following experiments, the action of each of these three general kinds of enzymes is demonstrated. As it is rather impractical to obtain enzymes from the frog, human and pig enzymes having the same action are used instead.

1. Action of a protease—pepsin.—

   a) *The gastric juice*: This digestive fluid is secreted by glands located in the wall of the stomach. It contains about 0.4 per cent hydrochloric acid, an enzyme called *pepsin*, and a number of salts. An "artificial" gastric juice is readily made by adding 0.4 per cent hydrochloric acid to commercial dried pepsin, generally obtained from the hog's stomach.

   b) *Action of pepsin:* Into a test tube put 5–10 c.c. of artificial gastric juice, and into another 5–10 c.c. of 0.4 per cent hydrochloric acid. Add to each a small quantity of boiled white of egg, cut into very small pieces. Place in a water bath or incubator kept at 37° C. (why?) for at least two hours. What becomes of the protein? Is the acid alone without the pepsin capable of producing this effect? Pepsin from the frog's stomach has the same action (Holmes, p. 142).

2. Action of a lipase—pancreatic lipase.—

   a) *Reaction of milk*: To 10 c.c. of milk in a test tube add a few drops of neutral litmus solution, or test with red and blue litmus paper. Litmus is a vegetable dye, which is pink in acid solution, blue in alkaline solution, and purplish in neutral solution. Is milk acid, alkaline, or neutral?

   b) *Action of pancreatic lipase:* A solution of this enzyme is obtained by dissolving dried pancreas, sold commercially as pancreatin, in a slightly alkaline solution. Add a few cubic centimeters of this solution to the litmus-containing milk prepared in (a). What color is the mixture? Keep in the water bath at 37° C., and observe from time to time. Does the color change? What does this indicate? The action is of course upon the cream (fat) of the milk (Holmes, p. 151). (R, L, A.) A similar lipase is secreted by the pancreas of the frog.

3. **Action of a diastase—ptyalin.**—

a) *Test for starch:* Place a few cubic centimeters of a 1 per cent starch paste in a test tube and add a drop or two of iodine solution. A deep blue color results, which proves the presence of starch. The nature of this blue compound which iodine forms with starch is unknown.

b) *Test for sugar:* Place a few cubic centimeters of glucose (fruit sugar) solution in a test tube, add a few drops of Fehling's solution, and heat to boiling. A yellow, greenish, or red precipitate proves the presence of sugar.

c) Test the starch paste with Fehling's solution. Is any sugar present in it?

d) Test some saliva, spit out from your mouth, with Fehling's solution. Does it contain any sugar?

e) *Action of saliva:* Put about 5 c.c. of the starch paste in a test tube, and spit some saliva from the mouth into it. Shake up thoroughly and place in the water bath at 37° C. for several minutes. Remove and test with Fehling's solution. Is sugar present? Where did it come from? The enzyme in the saliva which has this action is called *ptyalin* (from a Greek word meaning "spit"). An enzyme having a similar action is secreted by the pancreas of the frog (the mouth secretions of the frog have no digestive action).

Explain fully in your notebook the action of these enzymes. State your observations and interpret them. Consult Holmes, pp. 136–37, 142, 151. A general textbook of physiology, obtainable in the library, will also be helpful.

4. **Absorption.**—This is the process by means of which the products of digestion are transferred through the wall of the intestine into the blood and lymph vessels for transport to all parts of the body. The process is impractical of demonstration in the frog.

D. FUNCTION OF THE RESPIRATORY SYSTEM; RESPIRATION, OXIDATION

When any organic material burns, it uses up *oxygen* from the air and gives off *carbon dioxide*. In exactly the same way living substance burns either itself or the materials which it obtains from digestion, using up oxygen and releasing carbon dioxide. All of the processes involved in this burning are called collectively *respiration*.

1. **External respiration in the frog.**—The process and mechanisms involved in getting the oxygen of the air into the body and throwing out the carbon dioxide constitute external respiration. Obtain a live frog, place it in a covered dish with a small amount of water, and after it has become quiet study the respiratory movements. These consist of the opening and closure of the external nares,

rise and fall of the floor of the buccal cavity, and contraction and expansion of the sides of the body. *Time the rate of each of the movements.* Can you discover any correlation between any of these movements? Does the frog respire in the same manner as the human (*R, L, A*)? Read Holmes, pp. 168-177, understand thoroughly the mechanism of respiration in the frog and write an account of it in your notebook.

2. **Respiratory activity of the skin.**—Demonstration experiment. Three jars, each containing a frog, are placed in a trough of running water, so that the temperature of all is the same. One is filled with running water, the second with standing water, so that all bubbles of air are entirely excluded, and the third has an inch or two of water in the bottom. Do the frogs immersed in water show respiratory movements? How do they get oxygen? What is the condition of the three frogs after two or three days? **Explain fully** in your notebook.

3. **Carbon dioxide output.**—The following simple experiment demonstrates that carbon dioxide is given off from the lungs. (This need not be performed by those familiar with it.) A bottle is furnished containing two or three inches of lime water or saturated solution of barium hydroxide, and provided with two tubes, one of which extends below the lime water while the other does not. First draw atmospheric air through the lime water by sucking on the shorter tube. Does atmospheric air contain enough carbon dioxide to produce a precipitate with lime water? Then blow air from the lungs into the longer tube. What happens? Explain fully (*A*).

4. **Internal respiration or oxidation.**—The respiratory movements considered in sections 1 and 2 are merely methods for getting the oxygen into the body and into the blood. The actual use of the oxygen by the living substance of the animal is the real process of respiration. To avoid confusion, this process is designated as *internal respiration,* or *oxidation.*

E. FUNCTION OF THE EXCRETORY SYSTEM; EXCRETION

Every part of the organism as the result of its activities gives off waste matters. Those derived from the oxidation of fats and carbohydrates are mainly carbon dioxide and water, which are cast off through the lungs and skin. Those derived from the oxidation of proteins, or other chemical splittings of proteins, contain nitrogen, and these nitrogenous wastes are taken from the blood and lymph by the kidneys and eliminated from the body. This function of the kidneys is called *excretion.* The student should note that excretion is an active process, which is concerned with the elimination of soluble materials which have once been a part of the body; *therefore the purely passive ejection of undigestible materials from the intestine is NOT excretion,* and should not be confused with that process, as students are prone to do. Such undigested food in the intestine is designated as *feces,* and its passage from the body is the process of *defecation* or *egestion.*

FUNCTION OF THE CIRCULATORY SYSTEM; CIRCULATION OF THE BLOOD

The food prepared in the digestive tract and the oxygen taken in through the lungs must be carried to all parts of the body; and the waste materials from these parts conveyed to the kidneys and lungs. Such transport of materials is the function of the circulatory system.

Demonstration of the circulation of the blood. In the web of the frog's hind foot, spread out under the microscope, observe the following:

1. **The network of tubes,** the blood vessels, in which the blood flows.

2. **The composition of the blood.**—It consists of solid bodies, the *corpuscles*, which may be seen shooting along the vessels, and the colorless fluid, the *plasma*, in which they are suspended.

3. **The arteries,** vessels in which the blood flows from the larger vessels into the smaller branches.

4. **The veins,** vessels in which the blood flows from the smaller into the larger vessels.

5. **The capillaries,** the smallest vessels, forming a network, in which the direction of flow is indefinite.

6. **The pulse,** rhythmic jerks in the blood stream, due to the heart beats, discernible only in the arteries.

7. **The chromatophores,** the small black bodies in the skin. They may exhibit various shapes. In the contracted state, they are round black masses; in the expanded state, they show long, delicate, spidery processes.

FUNCTION OF THE REPRODUCTIVE SYSTEM; REPRODUCTION,
DEVELOPMENT, THE LIFE-CYCLE

The ovaries produce the female elements, which are called *eggs* or *ova*, and the testes produce the male elements, which are called *spermatozoa*. Before the egg can develop it must unite with a single spermatozoon, a process known as *fertilization*. The *fertilized* egg then undergoes a process of *development* which results in the production of an individual like the one from which the egg arose. The complete history from one individual to the next is called *ontogeny*, or the *life-cycle*. As the life-cycle of the frog occupies too great a period of time for its completion, the life-cycle of an insect will be studied instead. For this purpose, either the small fruit fly (*Drosophila*) or the common blowfly may be used. The latter is preferable owing to its larger size but is available only during the warm months. If *Drosophila* is to be used, each table will be given a bottle containing a pair of fruit flies and a piece of banana as food. If blowflies are to be used, a bottle will be provided containing two or three inches of moist sand; put a piece of liver in it and set it near an open window. Watch the gathering of flies about the bottle and observe, if you have time, the laying of eggs upon the liver by the female flies. Then stopper the bottle with a wad of cotton.

1. **The flies.**—Obtain a *Drosophila* killed by ether and note its characteristic insect structure: the division of the body into *head*, *thorax*, and *abdomen*, the characteristic rings or *segments* of which the abdomen is formed, the large eyes, three pairs of legs, and single pair of wings (most insects have two pairs of wings, but flies are characterized by one pair). Learn to distinguish male and female fruit flies (A). The males have slender rounded abdomens with a black area in the middle of the tip of the ventral side; the females have broader, more pointed abdomens, and the black markings are at the sides of the ventral surface of the abdomen. In male blowflies, the large eyes nearly meet in front while in the females there is a considerable distance between them. (If the experiment on heredity is to be performed, the two fruit flies given you will differ from each other in some striking way.)

2. **The eggs.**—Note and record the date on which the eggs are first observed. They are oval white objects, minute in the case of the fruit fly, much larger in the blowfly. Remove one and study under the low power of the microscope. Read carefully the directions regarding the use of the microscope, and consult the assistant on any points that you do not understand. The egg of the fruit fly has hexagonal sculpturings upon the surface which are said to be the impressions of the walls of the oviducts, and is provided with two oarlike processes which prevent the egg from sinking into the soft banana pulp, a circumstance which would probably be fatal to the larva when it emerges. The egg of the blow fly is marked similarly but less conspicuously with hexagons, has a concave surface which is the future dorsal side, and a convex surface which is ventral. Draw an egg of either fly.

3. **The larvae.**—Note and record the date on which the moving, wormlike *larvae* are first noticed. The larva develops inside the eggshell, and hatches forth rather suddenly by rupture of this shell. Remove a larva to a slide, anaesthetize with ether with the aid of the assistant, cover with a cover glass, and study under the low power of the microscope. Compare its structure with that of the parent fly. Is it more simple? Does it have head, eyes, wings, legs? Is the body divided into regions? Is the segmentation, or ringing of the body, more marked than in the parent? What animals do you know that are similarly ringed along their entire bodies? The significance of these facts may not yet be clear to the student, but they illustrate one of the most fundamental and general laws of development, that every organism in its development passed through stages simpler than itself, and stages that resemble animals lower in the scale of animal life than itself.

The most striking structure observable in the larva are the *tracheal tubes*, consisting of a pair of longitudinal trunks running the length of the body, ending posteriorly in a pair of large openings, anteriorly in a pair of smaller ones, and sending off extensive branches throughout the body. These tubes are full of air which enters them through the openings, called *spiracles*, and through their minute branches is conveyed to all parts of the body. This kind of air tubes is found only in insects and their near relatives and that of other animals where the oxygen is carried by the blood. The trusible end of the larva is a very imperfect head, with a mouth and pharynx which is provided with hard, black, curiously shaped jaws. readily be seen with the naked eye in the living larva and their action in the soft food should be noted. Each segment of the larva is armed with a ring of many small pointed teeth used to prevent slipping. The body larva is filled mainly by the digestive tract, which is much coiled and by the yellowish fat bodies, in which the excessive food taken in by the stored up to be used for the next stage in development. In the blowfly especially when they have become pretty large, there may be seen with naked eye a sac, the *food reservoir*, an outgrowth of the esophagus, filled the reddish liver which has been devoured. Make a drawing of the larva.

Note the remarkably rapid increase in size, or *growth*, of the larva. In the non-living food is transformed into the living substance of the organism (process of *assimilation*). What use do the blowfly larvae make of the meat? Which side of the meat do they frequent and why?

4. **The pupae.**—Record the date on which the motionless brown pupa first observed. The larvae, when fully grown, cease to feed, become motionless and shrink, leaving their skins to form the brown pupal cases. Within this case the adult fly develops from certain rudiments present in the larva, while most of the larval structure disintegrates and serves as food material for the developing adult. As time passes observe the adult structures appearing within the pupal case, the most conspicuous being the eyes and wings. Remove a pupa, examine under the low power and sketch. Is it segmented? In *Drosophila* the two prominent processes at the anterior end and the less conspicuous ones at the posterior end are the same spiracular openings noticed in the larva, and serve to admit air to the tracheal system. The pupa of the blowfly has no such projections, but the posterior end has a number of short spines of no evident function.

5. **The adults (imagines).**—Record the date on which adult flies are first noted. Observe that the pupal case is ruptured and left behind by the fly. How many days were required for the entire life-cycle? Would the time be the same for all periods of the year? How many flies were produced? Proportion of males and females? If all the offspring lived to produce at the same rate, how many would there be at the end of three months? What factors in nature prevent this enormous increase in numbers?

The life-cycle of a fly is a typical insect life-cycle. A sudden transformation in life-cycle, such as that from the larva to the adult, is called a *metamorphosis*. The transformation of the tadpole (polliwog) into the frog is another example of metamorphosis. On the other hand, many animals have no such sudden metamorphoses in their life-histories, but the development is slow and gradual. From what you have learned in this experiment, do you think that flies grow? Are small flies the young of larger flies?

H. SUMMARY OF PHYSIOLOGICAL PROCESSES

1. Food prepared in the digestive tract by the action of enzymes is absorbed through the walls of the intestine into the circulatory system.

2. Oxygen drawn into the lungs through the mechanical arrangement of the respiratory system passes through the walls of the lungs or through the skin into the circulatory system.

3. The circulatory system transports food and oxygen to all parts.

4. The living substance of the body withdraws the food and oxygen from the lymph and blood, and uses them:

a) For the formation of new chemical compounds or of new living substance (process of assimilation), thus accomplishing growth.

b) For the production of energy, by burning the food material with the aid of the oxygen (process of oxidation). The living substance itself may also be burned and produce energy.

5. The waste products resulting from the oxidation are excreted into the blood which carries them to the lungs, skin, and kidneys, where they are thrown out from the body. The lungs excrete mainly carbon dioxide and water; the skin, carbon dioxide, water, and dissolved waste matters; and the kidneys, dissolved nitrogenous waste materials.

6. The combined processes of assimilation, oxidation (or other energy-producing changes), and excretion are spoken of together as *metabolism*. Metabolism may be defined as the sum of those chemical changes taking place in protoplasm which result in the production of new compounds, new protoplasm, or of energy.

7. The energy produced in the metabolic processes is utilized to carry on the activities of the body, for the contraction of muscles, conduction of nerve impulses, secretion of digestive fluids, etc.

III. GENERAL HISTOLOGY: CELLS AND TISSUES

The *protoplasm* of which living bodies are composed does not exist as a continuous mass, but is divided up into minute portions, each of which is called a *cell*. A cell is defined as a small mass of protoplasm, containing a differentiated body, the *nucleus*. All organisms are either composed of a number of cells or consist of a single cell. That branch of biology which devotes itself to the detailed study of the different kinds of cells which occur in living things is called *histology*, and that part of histology which is concerned with the structure of the protoplasm in different kinds of cells is sometimes distinguished as *cytology*.

As most cells are of minute size, the microscope is necessary for their study. Read carefully the sheet of instructions regarding the use of the microscope, and consult the assistants on any points which you do not understand. Do not fail to heed the directions regarding adjustment of the light and use of the adjustment screws. Set up your microscope ready for use, take from your box of slides the slide labeled "*Necturus*—liver," and practice with this slide until you are familiar with the method of operation of the instrument.

NOTE.—As students frequently exhibit curiosity about the preparation of microscopic slides, such as are to be used in this section of the work, a word about this process may be introduced here. The piece of material of which slides are desired is removed from a freshly killed animal, placed in a fluid which kills it and preserves it in a nearly natural condition, hardened and dehydrated in alcohols of increasing concentration, and imbedded in some substance such as paraffin, which can be obtained in both liquid and solid condition. The paraffin containing the object is then hardened in a cold medium, cut into a rectangular shape and mounted on a machine called a microtome. In the microtome the paraffin block is moved up and down by means of an automatic micrometer screw across a very sharp knife edge, which slices off exceedingly thin sections of the material which is imbedded in the paraffin. Each such slice is called a section, and when these sections are mounted on the slide in the order in which they are cut, they are called serial sections. While such sections may be cut as thin as $\frac{1}{5000}$ of a millimeter, they are generally about $\frac{1}{100}$ of a millimeter. After the sections are mounted on the slide in such a way that they stick tightly, the paraffin is dissolved, and the piece of material is then stained in order to make the structures clearer by dyes which are similar to those used in dyeing cloth. Finally the sections are covered with a cover glass with the aid of some medium which cannot dry up.

24

The student will now understand that owing to the numerous processes and chemical agents to which the material is subjected in the making of a slide, artificial and abnormal appearances are frequently produced. Further, a difference in the angle at which the material is sliced will make two slides of the same material present a different appearance. In studying slides, the student must always bear these points in mind, and *should direct his attention only to those portions of the sections which are pronounced by the assistant to be typical and normal, and illustrative of the structures under consideration.*

A. STUDY OF A TYPICAL CELL

1. **Liver cells of *Necturus*.**—(*Necturus* is an amphibian, a relative of the frog, and often chosen for microscopic preparations because its cells are much larger than those of the frog.) Examine with the low power the slide marked " *Necturus*—liver" and note that the liver is composed of polygonal blocks. Each of the blocks is a cell, and we thus see that the liver is made up of such cells. Inspection of all parts of the animal would show that they, too, are similarly constructed.

Turn on the high power and examine the structure of a liver cell in detail (Hegner, pp. 26–29). Note:

a) The *cell wall*, or *cell membrane*, the delicate partition which separates each cell from its neighbors. In many cases, the cell walls may not be as distinct as they are here.

b) The *nucleus*, the spherical deeply stained body in the center of the cell. A membrane, the *nuclear membrane*, separates the nucleus from the surrounding *cytoplasm*. Within the nucleus the solid material takes the form of a network, the *linin network*, not very distinct here, on the fibers of which are strung the conspicuous, deeply stained irregular masses, the *chromatin granules*. This chromatin is a substance characteristic of and found with few exceptions only in the nucleus, and is recognizable by its staining properties. The meshes of the linin network are filled with a clear, transparent, invisible fluid, called the *nuclear sap*, the *nucleoplasm*, or *karyolymph*.

c) The *cytoplasm*, the portion of the protoplasm outside of the nucleus. Like the nucleus it consists of solid materials, imbedded in a transparent more fluid portion, variously known as the *ground substance*, or *cell sap*, or *hyaloplasm*. The solid material in these liver cells is apt to appear as a network, the apparent ´fibers of which are really rows of granules. This network in the cytoplasm is called the *spongioplasm*, and was formerly thought to represent the real structure of the cytoplasm; but it is now believed to be due to the action of killing fluids upon the protoplasm. Besides the spongioplasmic network, the cytoplasm frequently contains granules, droplets, fibers, etc.

Draw a cell under high power, showing all details.

2. Eggs of the sea urchin.—The eggs or ova of all animals are single cells. Examine with the low power the slide marked "*Arbacia*—mitosis" and note on it the sections of the eggs of the sea urchin (*Arbacia*). With the aid of the assistant find an unfertilized egg, recognized by its large, clear nucleus. Examine under the high power. Note the very large nucleus, containing an unusually large amount of nuclear sap, the chromatin granules in the nucleus, the large black spot in the nucleus, called the *nucleolus*, or *plasmosome*, the delicate nuclear membrane, and the cytoplasm packed with granules. The small black bodies clinging to the periphery of the egg are the male elements, or spermatozoa, each of which is also a single cell.

Draw, showing all details. The granular appearance is best imitated by stippling with the point of the pencil.

B. STUDIES OF TISSUES

The preceding study has been made on cells which are *generalized* in structure, that is to say, cells which are like those found in very simple animals and which have not become specialized for the performance of particular functions. In an adult organism so complex as the frog, however, there are very few cells which retain this elementary structure, but most of them depart from the type plan to a greater or less degree, depending upon the kind of function which they are called upon to perform. Moreover, for the better performance of these functions, the cells become united into orderly arrangements of layers or groups, and these associations of cells are held together and aided in the performance of their functions by certain materials, which they themselves secrete and which are called *intercellular substances.* Such an association of a number of cells of a particular kind with their particular kind of intercellular substance is called a *tissue.* We shall now study the various kinds of tissues (Holmes, chap. vi, pp. 121–33).

NOTE.—In these studies of tissues, *it is absolutely essential* that the student observe the following directions: (1) All material must be mounted in liquid, either water (if dead) or physiological salt solution (if living). It is absolutely useless to attempt to study dry material. The amount of liquid added should be just sufficient to come to the edge of the cover glass. (2) All material must be spread out as thin as possible, or picked into minute pieces with a pair of teasing needles. It is absolutely useless and a waste of time to try to study thick material. (3) After having made the material as fine as possible, put a' cover glass on, lowering it from one edge to avoid air bubbles. Excess fluid should be absorbed. The cover glass should not float around on the material. (4) Study the tissue first with the low power and without staining; then with the high power. Draw what you can see in the unstained material. (5) Then stain, if the directions say so, and add what you can see after staining to the drawing already made. Use a small amount of the stain and wait patiently for it to take effect. Much better results will be obtained than by piling on a lot of stain in an attempt to hurry matters. To stain, place a drop of the stain in contact with one edge of the cover glass, and draw it under by applying a piece of filter paper to the opposite edge of the cover glass. (6) If what you see in your preparation does not correspond with the description in the outline, then you should promptly conclude that you are looking at the wrong thing and should seek the assistance of the laboratory instructor. The descriptions in the outline have been made as accurate as possible. (7) Draw only a few cells, making your drawings large and detailed, putting in every structure that you can see.

1. **Epithelial tissues.**—This kind of tissue covers or lines all the free surfaces of the body, and is further distinguished by the relatively unspecialized character of its cells, which are similar in structure and appearance to the "typical" cells already described, and by the almost complete absence of intercellular substance. The cells are united into continuous sheets by a cement substance, which is difficult to demonstrate. There are several kinds of epithelia (Holmes, pp. 121-22).

a) *Squamous epithelium:* Obtain a small piece of shed epidermis (outer layer of the skin) of the frog, spread it out on a slide in a drop of water, cover with a cover glass, and examine with the low power. Turn down the light. Note the polygonal cells of which it is composed, giving a characteristic mosaic appearance; these cells are found to be very thin and flat when viewed from the side. Study a cell with the high power, note the nucleus and (in some cases) the pigment granules in the cytoplasm. If the nucleus is not clearly visible, stain with a drop of aceto-carmine. (See general directions for method of staining.) By changing the focus of the microscope determine whether the epidermis is one or more layers of cells thick. **Draw,** showing a few of the cells.

b) *Columnar epithelium:* In contrast to the preceding, this type of epithelium is characterized by the tall and slender shape of its cells. Obtain a small piece of the inner lining of the small intestine which has been macerated for twenty-four hours in 5 per cent chloral hydrate, add a few drops of salt solution, tear it into the smallest possible bits with a pair of teasing needles, cover and examine with the high power. Look for slender cells, slightly broader at one end and narrower or irregularly branched at the other. The oval nucleus occupies an enlargement which is usually nearer the narrow end. In some of the cells the broader end will be found to contain a cup-shaped cavity which in life is filled with mucus. Such cells are called *goblet cells.* **Draw** a few of the cells. If the nucleus is not visible, stain with a little aceto-carmine. In their natural position these cells form a single layer lining the cavity of the intestine, their long axes parallel to each other, and their broad ends facing the cavity.

Ciliated epithelium: Have the assistant give you a little tissue scraped from the roof of the mouth of a freshly pithed frog. Mount the scrapings in a little salt solution, tease into small bits, cover and examine. Search the cover until a shimmering movement is seen; then put on the high power. Note groups of cells with one surface covered with delicate hairlike processes of the cytoplasm, called *cilia*, which keep up a rapid vibration, sufficiently strong to move small particles in the vicinity or to cause the cells themselves to whirl about. You may be able to find some single cells. In that case note that while the cells in groups are more or less polyhedral in shape, with flat sides, the isolated cells tend to become spherical. These ciliated epithelial cells are arranged in a layer lining the roof of the mouth cavity, with their ciliated surfaces toward the cavity, and as the cilia of all the cells act in co-ordination, moving in waves which always travel in the same direction, a current is set up in that direction, as already demonstrated (experiment under II, B, 4). **Draw.**

2. Muscular tissue.—The cells composing muscular tissue are distinguished physiologically by their property of contractility, and morphologically by their long and slender form and fibrillar structure. As in the preceding kind of tissue, there is relatively little intercellular substance present (Holmes, pp. 128–31).

a) Involuntary, unstriated, or smooth muscle: Obtain a small piece of frog intestine which has been macerated, mount it in salt solution, tease with needles into the smallest possible bits, cover and examine. Notice that this material consists of very long and slender, almost threadlike cells, running parallel, in layers at right angles to each other. It is often difficult to actually isolate a single one of these long cells, but if you have teased your material carefully, you will generally find a few nearly separate cells along the frayed edges of the general mass. Each long fiber, which is an involuntary muscle cell, possesses an elliptical nucleus, usually faintly visible in the unstained material. If desired, the aceto-carmine stain may be applied to render the nuclei more distinct. **Draw.** (It may be remarked here for the benefit of the student that the figure in Holmes of these cells, Fig. 35, is somewhat inaccurate in respect to their length.)

b) Voluntary or striated muscle: Cut out a small piece from the leg muscles of your preserved frog or from a recently pithed frog (the former is often better for the purpose), mount in salt solution, tease carefully in the direction of the long axis of the muscle until you have separated your piece into threads, cover and examine. The long cylindrical objects which you will see are the muscle cells, or muscle fibers, and each voluntary muscle is composed of a great many such fibers. They are very large compared to the other cells which we have been studying, and in fact are so long that it is impossible to obtain a complete one; hence the ends are broken. Each fiber is covered by a delicate cell wall, called in this case the *sarcolemma*, and is crossed by conspicuous alternately light and dark transverse bands (really disks, as the muscle is cylindrical). It is because of these bands, which probably represent differences in the consistency of the muscle substance, that voluntary muscle is called "striped" muscle. A longitudinal striation may also usually be seen, indicating that the muscle fiber is really composed of a number of much smaller longitudinal fibers, known as *fibrillae* or *sarcostyles*, which are bound together in the same sarcolemma. Each muscle fiber contains a number of slender nuclei, located near the surface, just under the sarcolemma. In order to see them it will generally be necessary to apply the aceto-carmine stain. Since the young muscle cell possesses but a single nucleus, and since the others are produced by the division of this nucleus, it is probable that the adult muscle fiber is not a single cell but a number of cells with no cell walls between. Such a multinucleate structure is called a *syncytium*. **Draw** a muscle cell showing all structures.

3. **Connective tissue.**—In this type of tissue, the cells are much reduced and few in number, and the greater bulk of the tissue consists of intercellular substance. The function of the connective tissue is that of supporting and þinding other parts and tissues. It is therefore exceedingly widespread (Holmes, pp. 123–28).

a) *White fibrous or collagenous connective tissue:* This is the white weblike material binding the muscles of the frog together or forming the partitions between the subcutaneous lymph sacs. Remove some from either of these places of your preserved frog, spread it out carefully on a slide so as to make a very thin layer, add salt solution, cover and examine. It consists for the most part of long, slender, wavy fibers, running in all directions. These are the *white* or *collagenous fibers* (so called because when boiled they form glue); they are very tough and inelastic. Another type of fiber, the *yellow elastic fibers*, is present in small numbers, distinguishable by the fact that they run straight and singly, not in bundles as do the white ones. The fibers of connective tissue are not cells, and are probably not living, but they are the intercellular products of the real connective tissue cells. In order to see these latter, apply the aceto-carmine stain, and note after a few minutes the oval nuclei of the connective tissue cells staining deep pink or red, scattered among the fibers. The acetic acid in the stain will also tend to dissolve the white fibers, making the yellow elastic ones more distinct. Cells and fibers are imbedded in an invisible, clear, gelatinous *matrix* which is also secreted by the cells. **Draw** a small portion of the tissue, showing fibers and nuclei.

b) *Cartilage:* Have the assistant slice off with a razor a thin piece of cartilage from the end of the one of the long limb bones of a recently pithed frog, mount in salt solution, cover and examine. The prepared slide "Frog—cartilage" may also be used. Cartilage consists of a clear *matrix* which is much more dense and firm than in the previous type of connective tissue. In this matrix are rounded spaces or *lacunae*, at intervals, which are completely filled in the living material by the *cartilage cells*, which secrete the matrix. When two or more cells are contained in one lacuna, they have originated through the division of a previous single cell, and will become separated from each other by the ██████ between them. Draw a small portion of cartilage.

c) *Bone:* Examine prepared slide "Frog—bone." These slides ██████ by grinding down slices of dried bone; hence the bone cells and ███ structures of the bone have been destroyed. In bone, the matrix ██ rendered very firm and strong by the deposit in it of mineral salts, ██████ cium phosphate and calcium carbonate, through the activity of the bon██ This matrix is arranged in concentric layers, called *lamellae,* around ██ holes, which are the cross-sections of canals, the *Haversian canals.* The ██ sian canals traverse the bone in a longitudinal direction and in life car██████ vessels, lymph vessels, and nerves for the nutrition of the bone. Sca██████ through the lamellae are minute spaces, or *lacunae,* with spidery processes ██████ ing out into the matrix. In these spaces the *bone cells* and their processes a██ contained in the living condition. Draw a portion of bone.

4. **Blood.**—Blood may be regarded as tissue in which the intercellular sub██ stance is liquid. The cells of blood are called *corpuscles,* and the fluid inter██ cellular portion, the *plasma.* The corpuscles are of two general classes, *red corpuscles,* which give the red color to the blood and carry oxygen, and *white corpuscles,* which protect the body from disease (Holmes, pp. 258–64).

a) *Fresh blood:* Obtain from the assistant a drop of fresh frog blood, stir it up in salt solution, cover and examine with the high power. The numerous oval bodies are the red blood corpuscles, although they are not red except in masses. Scattered here and there among the red corpuscles will be found the white corpuscles, smaller, irregular in shape, and with granular cytoplasm. If you watch a white corpuscle for some time you may see it undergo slow changes of shape, an example of *amoeboid movement.* By gently moving the cover glass, cause the red corpuscles to float about on the slide, and as they turn determine their shape in profile. The central bulge is due to the nucleus, which is generally faintly visible.

b) *Stained blood:* Examine with the high power the slide of blood. Note:

(1) The red blood corpuscles, the very numerous oval bodies on the slide. Each has a central nucleus. Draw one.

(2) The white blood corpuscles: As these are much less numerous than the red corpuscles, it will be necessary to search the slide carefully for them. There are several kinds of white corpuscles. Those with a complex nucleus, consisting of several pieces, and with granular cytoplasm are called *leucocytes;* those with an ordinary type of nucleus and with clear cytoplasm are *lymphocytes.* Try to identify the following kinds and make drawings of those which you are able to find:

(a) *Polymorphonuclear* leucocytes: This type of white blood cell is distinguished by its very irregular nucleus, which consists of several masses, apparently separate but really united by delicate strands. The cytoplasm is always packed with granules, which are fine in the common type of leucocyte, coarse in a rarer type.

(*b*) Large lymphocytes: These are nearly spherical cells, with clear cytoplasm and a small rounded nucleus.

(*c*) Small lymphocytes: They are smaller than the preceding with relatively large nuclei, covered by a small rim of cytoplasm.

5. **Nervous tissue.**—Nervous tissue is composed of *nerve cells*, and the structures which support them (Holmes, pp. 131–33). A nerve cell consists of a central portion, called the *cell body* which contains a large nucleus, and of slender processes which extend out from this cell body often to long distances. Portions of nervous tissue which consist almost entirely of cell bodies are designated as *gray matter*, while those which consist of the processes constitute the *white matter*. The processes of a nerve cell are of two kinds: those that convey the impulses into the cell, called *dendrites*, usually very numerous and much branched, and that one which conveys the impulse away from the cell, always single and unbranched or slightly branched, named the *axone*. What are called *nerves* are bundles of axones.

a) *Brain cells:* In order to demonstrate the entire nerve cell with its processes it is necessary to cut thick slices of the brain, and to devise a special staining method since the processes do not take ordinary stains. A method was originated by an Italian histologist named Golgi, by which the entire cell is blackened by a deposit of silver upon it. Examine a slide of brain provided, stained by the Golgi method, either brain cells of the rabbit (*Lepus*), or human cerebellum. Use low power only, as the sections are thick. Note the numerous black branched objects upon the slide. Each is a nerve cell with its processes. Pick out a favorable place on the slide (*A*) and study one of the cells. Each consists of a rounded or triangular *cell body* from which spring several processes. In the case of the rabbit, one large, stout process springs from the pointed end of the cell body and several branched processes from the rounded end. In the cells of the cerebellum, two or three stout processes (dendrites) spring from one side of the cell and immediately break up into an exceedingly complicated system of branches, while from the opposite surface of the cell the small slender axone arises (it is not always visible). **Draw** a brain cell.

b) *Motor cells of the spinal cord of the frog:* In order to demonstrate the actual structure of the cell body, the usual thin sections, stained in the ordinary way, are employed. Examine the slide "Spinal cord—frog" with your lowest power. Identify the central large oval body on the slide (all other objects are to be disregarded). In this oval object, which is the cross-section of the spinal cord, observe two general regions, a central, denser, slightly darker region, the *gray matter*, trapezoidal in shape and containing numerous darkly stained cell bodies; and an outer region, lighter, more open, and with only a few small cell bodies, the *white matter*. In the two corners of the lower base of the trapezoid

.... by the gray matter beside several very large cells. These are the *motor* .... which passes out to the voluntary muscles of the body. Examine .... them with the high power. The structure of the cell body is well shown, ... but the .... of the processes are present. Why?) Study the .... in the cell wall, comparing with other cells already seen. Note the .... large nucleus, containing a conspicuous round body, the *nucleolus*, ... the .... granules on the granular, sometimes fibrillar structure .... Draw.

.... a small piece of fresh nerve from a recently .... it apart .... ease with needles in a longitudinal .... it into minute fibers. Cover and examine. .... number of cylindrical fibers, bound together .... fibers is in fact a single *axone* of a nerve cell. .... with the high power. It consists of a central .... often called here the *axis cylinder*, and a .... sheath known as the *medullary sheath*. The .... substance, called *myelin*, which tends to swell .... preparation. Outside of the medullary sheath .... Draw.

.... animals are the largest cells, although their .... large amount of inert food material (yolk) .... the frog are the spherical black and white .... are noted on the surface of the ovaries. .... of the frog are best studied in the spring, ... as at other seasons they are apt to be .... In such case, spermatozoa of a guinea- .... Take a small portion of the testis of a .... raise in salt solution, cover and examine .... will contain thousands of minute, slender, ... They are peculiarly modified cells. .... slightly curved, *head*, which is practically .... slender, vibratile *tail*, which represents .... animals generally have smaller .... of the frog. Draw a spermatozoön.

IV. GENERAL HISTOLOGY: STRUCTURE OF ORGANS

An *organ* consists of two or more kinds of tissue united together in a definite and characteristic way for the performance of specific functions.

A. STRUCTURE OF THE LIVER

Examine slide " *Necturus*—liver" (see Holmes, pp. 153–56). The liver may be taken as an example of a rather simply constructed organ, and also as an example of a *secreting gland*. It consists almost entirely of large cuboidal epithelial cells, arranged in cylindrical *columns* which branch and connect with each other in a very irregular manner. The columns therefore form a kind of network with numerous large spaces between them. These spaces are capillaries and in them blood corpuscles will usually be found. A very slender, usually invisible canal runs down the center of each liver column between the cells, and into this canal, which is called a *bile capillary*, the secretion of the liver cells, known as *bile*, is poured. The large black spots scattered abundantly through the liver are collections of *pigment* granules. Large blood vessels, and perhaps some of the *bile ducts* which collect the bile from the bile capillaries, may be present on the section. Surrounding the liver and penetrating it here and there along the course of the larger blood vessels is a small amount of white fibrous connective tissue. The liver cells have several other highly important functions besides that of secreting bile.

B. STRUCTURE OF THE INTESTINE

Examine slide " *Necturus*—intestine" (see Holmes, pp. 148–50). These cross-sections of the small intestine have been stained with Mallory's connective tissue stain in order to produce a marked color contrast between the different layers of the intestinal wall. This stain dyes connective tissue and related substances a deep blue color, and nuclei orange or red.

1. **General structure of the intestinal wall.**—The wall of the intestine consists of four coats. Identify these as follows with the low power, beginning next to the cavity:

a) The *mucous* coat (*mucous membrane* or *tunica mucosa*), the innermost light-colored layer, containing numerous red or orange nuclei. It is thrown up into folds and is sharply separated from the next layer by its different reaction to the dye.

b) The *submucous* coat (*tela submucosa*), a broad band, stained deep blue, containing frequent spaces and extending up into the folds of the mucosa.

... The .... the remainder of the intestinal ... .

... . *....... ....* the inner layer, in which the ... ... direction.

... . *....... ....* the outer layer, ... direction. This layer also contains . ... and hence takes more of the blue ....

... ... *.........* . a very thin mem- .... under the low power. It is .. .. investing the intestine.

a. *Detailed structure of the intestinal wall.*—Select a favorable place where .. .. area of its cells. The best .... Study with the high power.

.. .. lining *epithelium* of the intestine .. .. of a single layer of .. .. of the intestine to the under- ... ... These cells were studied in Section III, ... to see. Each cell possesses ... . stained red or orange. Why do the ... . all in the same plane (determine this . ... next the cavity) of many of the .. .. called *goblets*, which are filled . ... to connective tissue, the goblets .. .. There are thus two kinds of cells in . . .. and the *goblet cells.* Understand the . . .. the mucosa by considering that . ... be parallel to the long axes of . . . .

. ... with the epithelium and accompanying . ... a part of the mucous membrane. It is ... ., and consists of loosely woven white ... . is general parallel to the epithelium. The .... are distinctly visible) are more . . .. However, no sharp distinction can be . ... coat. The tunica propria . . .. spaces, and contains little nests . . ..

.. .. white fibrous connective tissue con- . ... is does do not run parallel to the cavity ... ... connective tissue cells and the many .. ..

c) The circular muscle layer consists of smooth muscle cells running parallel to the plane of section. The boundaries of the individual muscle cells are not clear, but the whole layer appears to be made up of parallel fibrillar striations. The conspicuous nuclei are long and spindle-shaped.

d) The longitudinal muscle layer contains a considerable quantity of white fibrous connective tissue which stains blue, in which are imbedded smooth muscle cells running at right angles to the plane of section, and thus appearing as circles. Understand why the nuclei may or may not appear in the circular cross-section. Numerous blood vessels are present in this layer.

e) The serous coat is a very thin membrane looking like a line closely applied to the outer surface of the preceding coat. At intervals its flattened nuclei are seen. As already explained the serous coat is the visceral peritoneum.

Draw a small portion of the intestinal wall showing the appearance and cell structure of the fiber layers accurately and in detail. The drawing should be four or five inches in width.

C. STRUCTURE OF THE STOMACH

Examine slide "Frog—stomach" (see Holmes, p. 140). Observe that the general appearance and arrangement of layers of the stomach wall are the same as in the case of the small intestine, except that the coats are thicker, especially the mucous and muscular coats. The particular point of interest about the stomach wall is the formation of *glands* in the mucosa.

1. **The gastric glands.**—The mucosa of the stomach is thrown up into regular folds or *rugae*, which appear as conical elevations in section. The lining epithelium does not form an even layer over these folds but is itself folded in and out in such a way as to produce a large number of long tubular glands set closely together with their long axes parallel. It may be noted in passing that all glands are produced by such infoldings of epithelia. Narrow strands of the tunica propria separate the glands from each other. Find a place on the slide where the glands are cut parallel to their long axes and examine one of the gastric glands in detail with the high power. (Their appearance is somewhat different in different regions of the stomach. Consult Holmes, p. 140.) The parts of a gastric gland are *mouth*, *neck*, and *body* or *fundus*. The mouth is the longest part and contains a narrow canal leading from the interior of the gland into the cavity of the stomach. The cells near the beginning of the mouth are very elongated, with clear free ends, probably containing mucus, and slender-tailed nuclei. Farther down the mouth, the cells and nuclei become more rounded. The region where the mouth ceases is the neck and consists of a few cells which contain large clear spaces. The gland may branch in the neck region so that two or more bodies are attached to one mouth. Below the neck the body or fundus is composed of polygonal granular cells, whose function is to secrete the gastric juice.

Draw a gastric gland in detail.

2. **The muscularis mucosae.**—Another difference between the stomach and the intestine is that the tunica propria of the mucous membrane is sharply bounded from the submucous coat by a narrow band of smooth muscle fibers which is called the *muscularis mucosae*. It is composed of the usual two layers, an inner circular and an outer longitudinal one.

D. STRUCTURE OF THE SKIN

Examine slide "Frog—skin" (see Holmes, chap. ix, pp. 179–86). The skin is a combination of epithelial and connective tissue. The epithelial part is called the *epidermis;* the connective tissue part, the *dermis* or *corium*.

1. **The epidermis.**—The epidermis is a *stratified* epithelium, i.e., an epithelium composed of several layers of cells. The outermost layer of the epidermis (*stratum corneum*) consists of very flat, thin, polygonal cells which are cornified or horny in composition. These cells are the ones which are shed by the frog, and they have already been examined separately. Beneath the stratum corneum, the cells gradually change from a flattened to a rounded, and finally to a columnar shape, until the innermost ones are quite columnar in form. These layers of cells constitute the *stratum germinativum* (also called *stratum mucosum*, and *stratum Malpighii*). The cells of the stratum germinativum frequently contain brown pigment granules, and dark brown or black branched *pigment cells* (*chromatophores*) may be scattered among the regular epithelial cells.

2. **The dermis.**—The dermis is composed of connective tissue, separable into two layers, an outer loose layer, the *stratum spongiosum*, and an inner compact layer, the *stratum compactum*.

a) The stratum spongiosum: This consists of a loose, irregular network of connective tissue fibers, inclosing lymph spaces, blood vessels, etc. It contains a thin layer of chromatophores, dark-colored, irregular, branching cells just under the epidermis, and beneath this the *cutaneous glands*.

b) The cutaneous glands: These are produced by a simple infolding of the stratum germinativum of the epidermis, and their walls are a single layer thick. Each opens to the surface of the skin by a narrow neck which passes up through the epidermis. Explain why most of the glands on the slide appear to have no neck (*A*). Also explain why some of them appear to be solid instead of hollow. Two general varieties of cutaneous glands are recognized in the frog, the *mucus* and the *poison* glands.

The mucus glands are smaller and much more numerous than the poison glands. Their appearance differs according to the stage of activity in which they happen to be. In the resting or inactive state, the epithelial cells are very high and conical so that the cavity of the gland is practically obliterated; the nuclei are small and situated at the bases of the cells. In the active state, the inner ends of the cells are converted into mucus which forms transparent masses in the cavity of the gland, which now appears larger than in the resting condition; the cells in the active condition are smaller, cuboidal, with centrally situated nuclei. Intermediate states, of course, occur between these two extremes.

The poison glands are much larger than the mucus glands and of infrequent occurrence except on certain parts of the skin. There may be none on your slide. The epithelial cells form a very thin layer surrounding the very large cavity, and frequently the cell walls are not distinct. The secretion, which is poisonous, appears as granular masses near the epithelium.

Each of the glands is invested with a thin layer of smooth muscles, and outside of this a layer of connective tissue. Neither of these layers can be clearly made out in the preparations.

c) The stratum compactum: This consists of a dense layer of white connective tissue fibers, arranged parallel to the surface of the skin. At intervals, vertical strands, consisting of connective tissue, smooth muscle cells, blood vessels, nerves, etc., cross the wavy layers of the stratum compactum at right angles, and may extend up into the epidermis.

d) The subcutaneous connective tissue: Beneath the stratum compactum occurs loose connective tissue which is not a part of the skin but forms the boundaries of the subcutaneous lymph sacs.

Draw a portion of the skin, showing in detail the structure of all the layers.

E. STRUCTURE OF THE KIDNEY

Examine slide "Frog—kidney" (see Holmes, pp. 202–6). The section is taken across the kidney. The flattened surface is ventral; the rounded surface dorsal. In the middle of the ventral surface, observe a yellowish or orange mass, the *adrenal gland*, made up of masses of epithelial cells (this is not present on all sections); and in this region also there are usually seen sections of large arteries and veins. The two slides of the kidney may be distinguished in section as follows: the outer edge is more pointed and contains cross-sections of a large vein, the renal portal vein, and the ureter.

1. **The tubules.**—The greater portion of the kidney is made up of extremely long tubular glands, called tubules, which are much coiled and twisted. Because of this coiling, the tubules appear in section as circles, ellipses, crescents, etc. Each tubule possesses a large and distinct central cavity, and its wall consists of a single layer of cuboidal epithelial cells, whose function is to extract the nitrogenous waste matters from the blood. Note the numerous blood vessels between the tubules, and the connective tissue between the tubules and covering the surface of the kidney.

2. **The Malpighian bodies or renal corpuscles.**—Each uriniferous tubule begins in a *Malpighian body*, little rounded masses situated near the ventral surface. Examine one of these with the high power. It consists of a dense tuft

of capillaries, called a *glomerulus*, around which the beginning of the tubule forms a very thin cup-shaped inclosing membrane, the *Bowman's capsule*. Through the opening of the cup an artery enters the glomerulus, and a vein leaves it; opposite this point the uriniferous tubule begins. You will probably have to examine a number of Malpighian bodies in order to see all of these points. The tubule after leaving the Malpighian body winds in a complicated way through the substance of the kidney and finally empties into a collecting tubule which opens into the ureter.

F. STRUCTURE OF THE SPINAL CORD

Examine slide "Frog—spinal cord" (see Holmes, p. 286). The spinal cord, as we have seen, consists of the central *gray matter*, composed of nerve-cell bodies, and the peripheral *white matter*, composed of nerve-cell processes. In the center of the gray matter is a small canal, the *central canal*. The four corners of the gray matter are produced into processes, known as the *dorsal* and *ventral horns*, or *cornua* of the gray matter. The ventral horns are considerably broader than the dorsal horns, and contain the large motor cells which have already been examined. The spinal cord is nearly separated into two symmetrical halves by the *dorsal* and *ventral fissures*, which extend from the middle of the dorsal and ventral sides in toward the central canal. The dorsal fissure is shallow and is continued toward the central canal by a narrow *septum*, composed of loose tissue. The ventral fissure is much deeper and extends nearly to the central canal; it incloses an artery, the ventral spinal artery.

The white matter consists mostly of cross-sections of the nerve fibers, which take their origin from the cells of the gray matter. In addition to the nerve fibers, the white matter contains scattered cells, which are mainly the cells of the *neuroglia*, a peculiar kind of connective tissue, found only in the nervous system. The neuroglia forms a network supporting the nervous structures.

The cord is surrounded by a sheath, the *pia mater*, composed of connective tissue, and containing numerous blood vessels.

Draw the section of the cord, showing the above-mentioned features. The instructor will be glad to loan sections of other organs of the frog to students who care to see them.

V. THE SPECIAL ANATOMY OF THE FROG

We have seen that organisms are composed of cells, that cells are combined to form tissues, and tissues to make up organs. The organs are united in groups called systems. We have already studied these systems in a general way, and they are now to be studied in detail.

A. THE DIGESTIVE SYSTEM

1. **Esophagus.**—Use your preserved frog. Cut through the tissues bounding the anterior wall of the coelome to the left of the left lung, and turn the lung and heart to the right. This exposes the *esophagus*. It extends from the end of the mouth cavity, called the *pharynx*, to the *stomach*. A slight enlargement usually marks the beginning of the stomach (Holmes, pp. 138–42).

2. **Stomach**—The end of the stomach continuous with the esophagus is called the *cardiac* end; the opposite end is the *pyloric* end, and is marked by a constriction, the *pylorus*. Slit open the stomach by a longitudinal slit from cardiac to pyloric end. Note the longitudinal ridges, or *rugae*, in the lining of the stomach, and compare with the lining of the esophagus and small intestine. Examine the cut surface, and note that the layers of the stomach wall, which have already been seen in microscopic section, can be distinguished with the naked eye. The mucosa and submucosa appear as a single layer, which can be separated from the underlying thick circular muscle layer; the longitudinal muscle layer forms a thin white outer coat.

3. **The small intestine.**—The first part of the small intestine is called *duodenum*, and it receives the *common bile duct* which carries the secretions of the liver and pancreas. The remaining coiled part of the small intestine is the *ileum* (Holmes, p. 148).

4. **The liver.**—The liver consists of the following parts: The *left lobe* is the largest, and is divided by a deep fissure into an *anterior* part, which lies to the left, and a *posterior* part which occupies the middle. The *right* lobe is not subdivided, but lies to the right of the median line extending somewhat dorsally. The *middle* lobe of the liver is not visible ventrally unless the other lobes are pushed apart. It then appears as a small squarish lobe extending far dorsally and continuous with the posterior part of the left lobe. Between the right lobe and the posterior part of the left lobe on the dorsal side is the green round *gall bladder* (Holmes, p. 152).

5. **The pancreas.**—Turn the lobes of the liver forward and study the pancreas. It is an irregular, branched, yellowish gland lying in the gastro-hepato-duodenal

ligament. It sends down two tail-like processes to the duodenum, and another process, the hepatic process, along the dorsal surface of the posterior part of the left lobe of the liver, where it extends almost to the gall bladder (Holmes, p. 131).

b. The bile duct.—The common bile duct runs through the pancreas and enters the duodenum by way of the second of the tail-like processes mentioned above, i.e., the one farthest away from the pylorus. Dissect away the pancreas in this region, and find the slender white duct, about the size of a sewing thread. Trace the duct up through the pancreas into the hepatic process of the pancreas. Here it receives a number of small hepatic ducts from the posterior part of the left lobe. To find these dissect away the substance of the liver. Follow the bile duct to the gall bladder, where it again receives hepatic ducts from the right lobe and the posterior part of the left lobe. Two ducts, the cystic ducts, emerge from the gall bladder. One of them unites with the hepatic ducts near by, and the other is continuous with the common bile duct. Consult Holmes, Fig. 42, p. 132, and work out the hepatic and cystic ducts as well as you can. It is not practical to find the pancreatic ducts.

c. The large intestine. The ileum enlarges abruptly into the large intestine, which runs straight to the anus. The upper part of the large intestine is called colon, the lower part, cloaca. Cut through the pelvic girdle so as to expose the intestine all the way to the anus. Find the origin of the urinary bladder from the ventral wall of the cloaca. The cloaca also receives the ureters and oviducts.

Draw the entire digestive tract, showing all its parts, with the lobes of the liver drawn correctly. Draw in the common bile duct in proper position in the pancreas and the hepatic and cystic ducts as far as you have been able to find them.

THE URINOGENITAL SYSTEM

Remove the digestive system, leaving the large intestine in place (Holmes, Figs. 35, 36).

a. The urinogenital system. Strip off the peritoneum wherever necessary to expose the urinogenital system. Each testis is an oval body attached by a fold of peritoneum, the mesorchium, to the adjacent kidney. From each testis arise a number of delicate ducts, the vasa efferentia, which run in the mesorchium toward the kidney. The vasa efferentia are really outgrowths of the kidney tubules which branch the kidney, which extend out and connect with the testes. By holding up the mesorchium to the light, the vasa efferentia may usually be seen. The vasa efferentia after entering the kidney eventually connect with the ureter, which differs in different species of frogs. From the posterior, ventral part of each kidney extends the ureter. Find it and trace it to the cloaca. Near and parallel to the lateral edge of the kidney runs a vestigial *oviduct*, which also opens into the cloaca. Trace it to the cloaca and note that the ureter is closely applied to the dorsal side of the oviduct.

Draw the male urinogenital system showing all of the above-mentioned parts.

2. **Female urinogenital system.**—The relations of kidney and ureter are the same as in the male. The ovary, however, bears no relation to the kidney. It and its mesentery have already been noted. The oviducts are a pair of convoluted tubes, extending the whole length of the body cavity. Trace one anteriorly to the anterior wall of the coelome, where it opens into the body cavity by a funnel-shaped mouth, or *ostium*, near the base of the lung. Have the assistant demonstrate the ostium to you. Trace the oviduct posteriorly to the cloaca. It widens into a large, thin-walled sac, the *uterus*, which lies behind (dorsal to) the peritoneum in the cisterna magna. Cut open the cloaca, and locate in its dorsal wall the two openings of the oviducts, situated upon projecting papillae and in the ventral wall, the opening of the urinary bladder.

Draw the female urinogenital system with the cloaca cut open, showing all parts.

C. THE RESPIRATORY SYSTEM

At their anterior ends the two lungs open into a chamber, the *larynx*, which communicates with the pharynx through the slitlike *glottis*. The larynx lies just in front of the heart. Consult Holmes, chapter viii, pp. 165–68. Dissect away all the muscles from the under surface of the lower jaw so as to expose the flat body of the hyoid cartilage. Find the two *thyroid processes* of the hyoid which extend posteriorly and inclose the *laryngeal chamber* between them. The walls of this chamber are supported by a complicated arrangement of cartilages for which Holmes, Fig. 45 (p. 166), should be consulted. Dissect posteriorly from the larynx and remove or cut through blood vessels or other structures until you have exposed the connection of the lungs with the larynx. Make a longitudinal slit through the ventral wall of the larynx, spread the slides apart and look within for the *vocal cords*, a pair of folds extending lengthwise in the chamber. Find the openings of the lungs into the latero-posterior walls of the larynx, insert one blade of a scissors into one of the openings and slit open the wall of the lung. Note that the inner wall of the lung is raised up into a network of ridges, which divide the wall into a large number of small chambers or *alveoli*. Blood vessels run along the ridges and break up into an intricate network of capillaries in the walls of each alveolus, which serves as an air sac.

Draw the dissection.

D. THE CIRCULATORY SYSTEM: THE VENOUS SYSTEM

For this purpose a fresh frog, probably injected, will be supplied. Open it up in the usual way to the left of the median line, but be very careful not to cut any blood vessels, especially near the heart. Be very cautious in spreading the body walls apart near the heart, so as not to injure the veins. The anterior abdominal vein in the median ventral line *must be preserved* and should be separated from the body wall before spreading the walls apart.

1. **The systemic veins** (Holmes, pp. 272–74).—*Systemic veins* are those which open into the heart, returning blood from the body directly to the heart. Remove the pericardial sac from the heart, and turn the heart forward so as to expose the sinus venosus. Three large veins contribute to the sinus venosus: the *posterior vena cava*, or *postcaval* vein, which emerges from the liver (receiving a large vein on each side from the liver so that apparently three veins are present) and enters the median posterior wall of the sinus; and the two *anterior venae cavae*, or *precaval* veins, one opening into each lateral wall of the sinus.

a) *The precaval vein:* Trace one of the precaval veins (since both have identical branches). It extends laterally from the heart along the border of the auricle, and about 1 cm. away from the heart passes through the pleuroperitoneal membrane. Directly outside of this membrane it forks simultaneously into three branches. The most anterior one of these is the *external jugular;* it passes straight anteriorly into the muscles of the floor of the mouth, from which, together with the tongue, hyoid, etc., it receives venous blood through numerous small branches. The middle of the three branches of the precaval is the *inominate;* it passes laterally and then turns abruptly dorsally, receiving as it turns a small *subscapular* vein from the muscles of the shoulder. The main vein beyond the entrance of this branch is known as the *internal jugular;* it descends straight into the hollow between the fore limb and the larynx, and disappears dorsally where it collects blood from the brain and spinal cord, from the roof of the mouth, and from a number of muscles. The third and posterior branch of the precaval is the *subclavian.* It is a large vein passing directly laterally to the base of the fore limb where it receives the *brachial* vein, carrying venous blood out of the fore limb. Just beyond this point the vein, now called *cutaneous*, or *musculo-cutaneous*, turns abruptly and runs straight posteriorly along the muscle of the ventro-lateral body wall (pectoral muscle). It then bends sharply dorsally and passes to the skin by way of the partition between the abdominal and lateral lymph sacs. It runs anteriorly under the skin forward to the nares, collecting venous blood from the skin of the entire dorsal side.

b) *The postcaval vein:* Trace the postcaval posteriorly from the sinus venosus into the liver. Where it emerges from the liver substance it receives two large *hepatic* veins, one from the right lobe, the other from the left lobe of the liver. Turn the lobes of the liver to the right and find where the postcaval enters the middle lobe from behind. Trace it posteriorly. It originates between the two kidneys, from which it receives a number of *renal* veins as well as veins from the reproductive organs and fat bodies.

2. The hepatic portal system (Holmes, p. 277).—A *portal system* is one in which the venous blood does not return directly to the heart but enters a system of capillaries in some organ from which the venous blood is then collected by a systemic vein. There are two portal systems in the frog, the *hepatic* portal system in which the interposed capillaries are in the liver, and the *renal* portal system where they are in the kidney. Spread out the liver with its lobes turned forward so that the pancreas is visible. Find a large vein ascending through the substance of the pancreas and forking on the dorsal surface of the liver. This is the *hepatic portal* vein which breaks up into a system of capillaries in the liver. The left fork of the hepatic portal vein sinks into the substance of the posterior part of the left lobe of the liver. The right fork connects with the anterior abdominal vein which also forks and then penetrates the liver. Trace the hepatic portal vein posteriorly and note that it is formed by the union of veins from the stomach, pancreas, spleen, and small and large intestines. All of the blood from the digestive tract therefore enters a capillary system in the liver, from which it flows into the postcaval vein by way of the hepatic veins. What is the purpose of this arrangement? Consult Holmes, p. 153.

3. The renal portal system.—Expose the kidneys and find along the outer lateral edge of each a conspicuous vein, the *renal portal* vein, which enters the substance of the kidney. Trace the renal portal vein posteriorly. At about the middle of the kidney it receives a vein, the *dorso-lumbar* vein, from the muscles of the back. Posterior to the kidney it is seen to be formed from two veins which come from the leg. The outer of the two veins is the *femoral*, the inner, the *sciatic*. At the point where each femoral vein enters the coelome, it gives off a branch, the *pelvic* vein, which runs along the posterior wall of the coelome to the median line where it joins the other pelvic. This union of the two pelvic veins produces the abdominal vein, which joins the hepatic portal vein as described above. The abdominal vein is thus a connection between the renal and hepatic portal systems, and blood from the hind legs may return to the heart either by way of the kidneys or by way of the liver. In the kidneys the renal portal vein breaks up into capillaries, from which the blood is collected by the postcaval vein.

4. The pulmonary veins.—From each lung a *pulmonary* vein passes dorsal to the sinus venosus and enters the left auricle. These veins are sometimes difficult to locate. Turn the heart forward holding it down with the finger so that it will be stretched. Pull away with a forceps the coronary ligament of the liver so as to expose the lungs. A short straight vessel will be found running from the inner border of the base of each lung obliquely toward the sinus venosus. These two pulmonary veins join just under the beginning of the left precaval vein and penetrate the left auricle at that point.

Make an outline drawing of the frog and its organs, and in this put the *venous system* in its proper relations to the organs. More than one drawing may be necessary. Draw only those vessels which you have found.

The conus arteriosus starts from the base of the right side of the ventricle, passes obliquely across the auricles, and divides near the anterior border of the auricles into two vessels which turn respectively to the right and left. Each branch is called a *truncus arteriosus* and each gives rise shortly to three arteries, which are designated as *aortic* or *arterial arches*. To find this division into three, carefully pick off the connective tissue, etc., from the surface of one truncus, and follow it away from the heart. The most anterior of the three arches is the *carotid arch;* the middle one is the *systemic arch;* the most posterior, the *pulmo-cutaneous arch*. The pulmo-cutaneous arch generally branches off before the other two (Holmes, pp. 268–70).

1. **The carotid arch.**—This vessel passes forward and soon divides into two branches, a medial small *external carotid*, and a lateral larger *internal carotid*. Just at the place of division, or situated on the internal carotid, is an enlargement, the *carotid gland*, presenting a blackish appearance owing to the presence of pigment cells. The internal carotid supplies the roof of the mouth, eye, brain, and spinal cord. The external carotid (or *lingual*) supplies the tongue, thyroid, and various muscles. Trace the branches of the carotid arch as far as practicable.

2. **The pulmo-cutaneous arch.**—Trace this out from the truncus arteriosus, and note that it soon divides into two branches. One of these, the *pulmonary*, is short, and runs directly to the lung. The other, the *cutaneous*, passes outward and forward, crossing the systemic arch, and then by a sharp dorsal turn disappears in front of the shoulder. Turn the frog dorsal side up, slit up the skin of the back, and deflect it. Find where the cutaneous artery emerges just in front of the suprascapula, and note its branches on the skin.

3. **The systemic arch.**—The systemic arch passes laterally and forward, and seems to disappear in the mass of muscles. Dissect away these muscles and follow its course. It runs forward alongside the internal carotid, then bends dorsally and turns backward dorsal to the esophagus. Just after it has made this posterior turn, the systemic arch gives rise to the following branches practically simultaneously: the small *esophageal* arteries, which supply the esophagus; the *subclavian*, which passes laterally to the foreleg; and the *occipito-vertebral*, which passes dorsally and promptly divides into an anterior branch, the *occipital*, and a posterior branch, the *vertebral*. Both of these can be readily followed in well-injected specimens. Turn the frog dorsal side up, and find the occipital artery emerging just under the anterior border of the suprascapula. Follow it along the head, noting a branch above the eye to the

anterior parts of the head, and one behind the eye for the upper and lower jaws. Remove all muscles from the vertebral column, and note the intricate branches of the vertebral artery to the vertebrae.

The two systemic arches from the two sides of the body now converge and unite to form a single large vessel, the *dorsal aorta*. The dorsal aorta lies in the cisterna magna, above the kidneys, and the pleuroperitoneum must be broken through in the usual way to one side of one of the kidneys in order to follow this vessel. At the point of union of the systemic arches, the large *coeliaco-mesenteric* artery arises. Trace the branches of this artery. It divides into a *coeliac*, which supplies the stomach, pancreas, and liver (*right* and *left gastric* arteries, and *hepatic* artery), and an *anterior mesenteric*, which gives off a short branch to the spleen (*lienal* artery), a number of branches to the small intestine (*intestinal* arteries), and one or more branches to the large intestine (*anterior haemorrhoidal* arteries).

Posterior to the origin of the coeliaco-mesenteric, the dorsal aorta gives rise to a number of small paired *urinogenital* arteries, which supply mainly the kidneys, but branch also to the fat bodies, and reproductive organs and their ducts. In the same region, one to four pairs of small *lumbar* arteries arise from the aorta and pass to the dorsal body wall.

The dorsal aorta next gives off a single median branch, the *posterior mesenteric*, to the large intestine, and shortly after this bifurcates into two large arteries, the *common iliac* arteries. Each common iliac soon gives rise to two branches. One of these, the *epigastrico-vesical* artery, branches almost immediately into an *epigastric* artery for the muscles of the lateral and ventral abdominal walls, and a *recto-vesical* artery for the urinary bladder and the rectum. The other is a small artery to the urinogenital ducts. The iliac then supplies a small *femoral* artery for the muscles of the thigh and continues down the leg as a large vessel, the *sciatic* artery.

Make an outline of the frog and its organs and put in the *arteries*, showing their proper location and distribution. All of the arteries mentioned can be readily found in well-injected specimens. Omit those that you could not identify.

Be able to trace the blood from any part to any other part.

F. THE CIRCULATORY SYSTEM: THE STRUCTURE OF THE HEART

The four parts of the heart, the sinus venosus, the auricles, the ventricle, and the conus arteriosus, have already been noted. **Draw the heart** from the side; show these parts.

The dissection of the heart should not be attempted except upon a well-preserved, uninjured heart. The heart of injected specimens cannot be used (Holmes, pp. 64–68). With a fine scissors remove the ventral wall of the ventricle, the ventral wall of each auricle, and the ventral wall of the conus arteriosus as far as the forking of the trunci arteriosi. If desirable the heart may be removed from the body by cutting through the aortic arches and the sinus venosus; then mount the heart under water in a small wax-bottomed dish. Note the thick spongy walls of the ventricle and its small central cavity. Find the openings of the two auricles into the ventricle. The walls of the auricles are quite thin, and the right one is considerably larger than the left. The large opening of the sinus venosus into the right auricle is readily found. The conus arteriosus contains a conspicuous structure, the *spiral valve*, which almost fills its cavity. It is a fold, bent slightly into an S-shape, extending lengthwise in the conus, its dorsal edge attached along the entire extent of the dorsal wall of the conus, its ventral margin free. In the cross-section of the truncus arteriosus observe the two partitions which divide it into three channels, one for each aortic arch. The ventral channel leads to the carotid arch, and the middle one to the systemic arch; the beginnings of these two channels are in the conus arteriosus in front of the termination of the spiral valve, so that the blood to reach them must flow over the cuplike widened anterior end of the valve. The dorsal channel of the truncus arteriosus passes into the pulmo-cutaneous arch; it starts farther down in the conus below the anterior end of the spiral valve, so that blood reaches it by flowing over the free ventral edge of the valve. Read Holmes, pp. 277-79, and understand the function of the spiral valve and the partitions in the truncus arteriosus in directing the venous blood into the pulmo-cutaneous arch, mixed blood into the systemic arches, and the arterial blood into the carotid arch.

Draw the dissection from the ventral side.

G. THE NERVOUS SYSTEM

Remove the skin and clean away all muscles from the median dorsal region of the animal. With a fine scissors and forceps remove the roof of the skull bit by bit, being careful to keep the point of the scissors well up against the bone so as not to jab into the soft brain tissue. First cut into the bone, then pull away the pieces with the forceps. Remove the roof as far forward as the nares. Next work posteriorly in the same way. Cut through the neural arches of the vertebrae by lateral cuts, and pull out the central piece with the forceps. The brain and cord should be well exposed before their study is undertaken (Holmes, chap. xvi, pp. 283-95).

1. Dorsal aspect of the brain (Holmes, pp. 291-95).—The brain is covered with a pigmented membrane, the *pia mater*, which is particularly abundant in the posterior part of the brain, where it is very vascular and fills a triangular cavity, the *fourth ventricle*. The pia mater should be removed from the fourth ventricle.

The most anterior part of the brain comprises a pair of rounded *olfactory lobes*, which are separated from each other by a faint median groove. Each olfactory lobe continues forward to the nasal sac as the *olfactory nerve*. The posterior boundary of the olfactory lobes is marked by a transverse groove which separates them from the next part of the brain, a pair of long oval bodies, the *cerebral hemispheres*. Just behind the cerebral hemispheres is a depressed region, the *diencephalon* or *thalamencephalon*, from which a delicate stalklike body, the *pineal body*, ascends dorsally to the brow spot. The pineal body is almost always torn off in removing the skull and hence cannot be seen. Posterior to the diencephalon are the two rounded *optic lobes*. Just behind them and forming the anterior wall of the triangular cavity of the fourth ventricle is the *cerebellum*. This is much smaller in the frog than in most other vertebrates. The region of the brain posterior to the cerebellum, and forming the floor and lateral walls of the fourth ventricle is known as the *medulla oblongata*. It narrows posteriorly until it becomes of the same width as the spinal cord with which it is continuous.

2. **The cranial nerves** (Holmes, p. 295).—Ten pairs of nerves spring from the lateral and ventral surfaces of the brain but most of these cannot be found unless very great care is exercised in dissecting. The first pair of cranial nerves, the olfactory, has already been noted arising from the olfactory lobes. The second pair of nerves, the *optic* nerves, springs from the ventral surface of the diencephalon, and may be seen by gently pushing this part of the brain to one side. Each penetrates the adjacent eyeball where it forms the retina. The third, fourth, and sixth nerves are motor nerves to the muscles of the eyeball and are too small to be found. The fifth, seventh, and eighth nerves arise close together from the side of the anterior end of the medulla, and may usually be seen by gently pressing back this region. The ninth and tenth arise together from the side of the medulla a short distance behind the eighth.

3. **The spinal cord.**—The spinal cord is continuous with the medulla oblongata and occupies a cavity within the vertebral column known as the *neural canal*. Posteriorly the spinal cord tapers into a fine thread, the *filum terminale*, which occupies the cavity of the urostyle and can be seen by cutting off the dorsal half of the urostyle. The spinal cord is slightly enlarged opposite the fore limb (*brachial enlargement*) and again anterior to the filum terminale (*sciatic enlargement*). These swellings are, of course, caused by the origin of the nerves to the limbs at those regions. In the median dorsal line of the cord is a longitudinal groove, the *dorsal fissure*. In a well-dissected specimen, the *dorsal roots* of the *spinal nerves* may be seen arising at regular intervals from the dorso-lateral region of the cord. The posterior roots run within the neural canal for a little distance alongside the spinal cord before passing out of the vertebral column.

Draw the brain and cord from the dorsal side.

4. **Ventral aspect of the brain.**—Cut across the cord back of the medulla oblongata, and carefully remove the brain, leaving the cord in place. In removing the brain note and cut through the more conspicuous of the cranial nerves. Study the ventral surface of the brain and identify the parts already noted on

he dorsal side. On the ventral side of the diencephalon observe the crossing of the optic nerves after their origin from the diencephalon: this crossing is called the *optic chiasma*. Just behind the optic chiasma is a bilobed extension of the floor of the diencephalon, called the *inferior lobes* or *infundibulum*; and to this is attached a rounded glandular body, the *hypophysis*, which fits into a depression in the floor of the skull, and is usually, therefore, torn off in removing the brain. The hypophysis is a gland of internal secretion.

Draw the ventral surface of the brain.

5. **The ventricles of the brain.**—The brain like the spinal cord is hollow, its cavity being continuous with the central canal of the cord. The cavities of the brain are known as *ventricles*. Make a median sagittal section of the brain, float it under water, and identify the ventricles. See also Holmes, Fig. 83 p. 202, and Fig. 84, p. 203. The cavity of the medulla oblongata is the largest and most posterior of the ventricles; it is called the *fourth* ventricle and has a thin vascular roof which has already been removed. From the fourth ventricle a narrow passage, the *iter* or *aquaduct of Sylvius*, extends forward below the optic lobes. Each optic lobe has a cavity, the *optic* ventricle. The ventricle of the diencephalon which extends downward into the infundibulum is the *third* ventricle. The *first* and *second* ventricles, also called the *lateral* ventricles, are inside of the cerebral hemispheres, which should be cut open to see them. The narrow passage which connects the lateral with the third ventricles is known as the *oramen of Monro*.

6. **The spinal nerves** (Holmes, pp. 89–91).—Turn the frog ventral side up, and remove all the viscera. Observe the spinal nerves passing out symmetrically from the sides of the vertebral column. They arise from the cord and leave the neural canal by way of openings (*intervertebral foramina*) between the vertebrae. There are ten pairs of spinal nerves, each of which divides immediately into a small *dorsal branch* and a larger *ventral branch*. It is the ventral branch which one sees running on the inside of the dorsal body wall. At the point of exit of each spinal nerve from the intervertebral foramen is a white mass, the *calcareous body*, which surrounds and conceals a *ganglion*.

Identity and observe the course of each of the spinal nerves. The first is quite small, it arises between the first and second vertebrae and innervates the tongue and muscles of the hyoid. The second is a large stout nerve which innervates the muscles of the fore limb. It is joined by branches of the first and third spinal nerves, and all of these together form a network which is called the *brachial plexus*. The presence of such a plexus indicates the compound origin of the muscles of the limb. The fourth, fifth, and sixth nerves are small and pass somewhat obliquely backward to supply the skin and muscles of the body wall. The seventh, eighth, ninth, and tenth nerves arise close together, run almost directly backward, and are united with one another by cross branches to form the *sciatic plexus* from which nerves go to the muscles and skin of the hind limbs. The small tenth nerve arises from the urostyle, and innervates mainly the cloaca, and urinary bladder.

Draw the spinal nerves as seen on the dorsal body wall.

7. The roots of the spinal nerves.—Select one of the largest spinal nerves (as the eighth), and trace it carefully into the vertebral column, cutting away the vertebrae. Pull off the calcareous body and find within it a small brown object, the *dorsal* or *spinal ganglion*. (The term "ganglion" means a mass of nerve cell bodies lying outside of the brain and spinal cord.) Tracing the nerve farther in toward the cord note that it divides into two branches or *roots*, one of which is attached to the dorso-lateral region of the cord, the other to the ventro-lateral region. These roots are designated as the *dorsal* and *ventral roots*. The dorsal root springs from the dorsal horn of the gray matter of the cord; it carries sensory fibers which arise from the nerve cells located in the dorsal ganglion. The ventral root takes its origin from the ventral horn of the gray matter, from the large motor cells which have already been seen in that location, and carries motor fibers to the muscles. The two roots meet just beyond the spinal ganglion, which is on the dorsal root, and the spinal nerve thus formed soon divides into dorsal and ventral branches or *rami*, as noted in the preceding section. Each ramus carries both sensory and motor fibers and supplies both skin and muscles; and the ventral ramus is further connected by the *ramus communicans* with the sympathetic system.

Make a diagram to show the origin of a spinal nerve from the cord.

8. General remarks on the function of the brain and cord.—The possibility that any organism born into the conditions of life as they exist upon the earth will survive depends entirely upon its ability to perceive and respond effectively to those conditions. We have already found that this capacity for the perception of conditions in the environment and for responding to them is vested in the nervous system. The perceiving part of the apparatus is the sense organs, which comprise the eye, ear, and nose, taste organs in the mouth, organs of touch, pressure, pain, temperature, chemical sense, etc., in the skin, and sensory organs in the viscera. The responding part of the apparatus is the brain and cord through the motor nerves to muscles and glands.

In a prone animal moving with one end forward, that end will come first in contact with the factors of the environment and hence will naturally come to be the place where the most important and specialized sense organs are located. Further, the part of the nervous system connected with these important sense organs must become enlarged to accommodate the numerous impulses sent in from them, and must thus acquire dominance over the rest of the central nervous system. We may thus account for the origin of the head and brain, which structures indeed made their appearance in the simplest bilateral animals.

We therefore find that the brain of a relatively simple vertebrate like the frog consists in large part of centers for the reception of the chief sensations.

Thus the olfactory lobes, the larger part of the cerebral hemispheres, and the dorsal and ventral parts of the diencephalon are olfactory. The lateral walls of the diencephalon and part of the optic lobes are the receptive regions for vision. Another part of the optic lobes and the dorsal margins of the anterior end of the medulla are the centers of hearing. The general sensations from the skin of the body are received into the dorsal regions of the spinal cord, which extend forward to the same regions of the medulla, where similar sensations from the head also enter. The dorsal and lateral regions of the medulla, therefore, together with the same regions of the optic lobes and diencephalon are centers of general sensations. Taste and other visceral sensations are received in the ventro-lateral regions of the medulla. The ventral portions of the brain from the optic lobes posteriorly and down the whole length of the cord in the ventral horns of the gray matter are the regions of origin of motor impulses.

Obviously these primary sense centers must be connected with the motor nerve cells from which nerves go to the muscles in order that an appropriate response may be made to the conditions in the world outside which are reported to the brain. Thus all the sense centers form intricate connections with motor centers for the production of motor actions. The spinal cord and part of the medulla are pathways for sensations from the body below the head to reach the brain and for motor impulses to reach the body muscles. They also carry out many reflex actions, i.e., actions resulting from direct connections of sensory impulses with motor responses without the aid of the brain. The medulla is further an important center of visceral functions, such as respiration, heart-beat, etc.

A still further mechanism is, however, required. This is a mechanism for the association and correlation of sensory information and for deciding between simultaneous ones. Thus suppose an animal smells some food and sees an enemy simultaneously. It must make a choice between the motor reactions which would result from each of these sensations if they were received separately. It must act "intelligently" in such situations. Such centers of correlation are naturally poorly developed in simple animals and become more and more prominent in the brain as the complexity of the animal increases. In the frog correlation is affected mainly in the optic lobes and diencephalon and to a slight extent in the cerebral hemispheres. Hence the complete removal of the cerebral hemispheres in the frog is of little consequence to the animal, except that the sense of smell is lost (see Holmes, p. 309). In higher vertebrates the cerebral hemispheres become more and more important as the seat of correlation, co-ordination, and intelligent action.

The cerebellum is also a co-ordination center, but one not associated with consciousness. It co-ordinates, reinforces, and exercises general control over motor movements, including the maintenance of equilibrium.

For this work dried mounted skeletons will be provided. (The student may prepare a skeleton as follows: Remove all possible flesh and organs from a freshly killed frog, and dip what remains from time to time in hot water, brushing away the remaining flesh until the bones are cleaned. Too liberal use of hot water or boiling will cause the bones to fall apart.)

1. **General considerations on the skeleton.**—The skeleton, or hard parts of the body, is generally of two kinds: the external skeleton, or *exoskeleton*, and the internal skeleton, or *endoskeleton*. The exoskeleton covers and protects the body and is derived from the skin. Examples are scales, feathers, hair. The frog has no exoskeleton except such as has become fused to the endoskeleton (see below). The endoskeleton is the internal framework of the body, and is composed of cartilage in the embryo which becomes partially converted into bone in the adult frog. Such bone formed in cartilage is known as *cartilage bone*. Investigation shows that not all the bones of the frog's skeleton have arisen in this way but some have appeared in connective tissue without passing through a cartilage stage. Such bones are called *membrane bones, dermal bones*, or *investing bones*. They are produced from the dermis of the skin, and are therefore not really endoskeleton, but part of the exoskeleton. In the course of evolution these bones, originally scales, sank into the body and attached themselves so closely to the endoskeleton that they are usually treated in textbooks as parts of the endoskeleton. The clavicle and the superficial bones of the skull and jaws of the frog are membrane bones; all other bones of the skeleton are cartilage bones.

The skeleton may be divided into an *axial* part including the *skull*, the *visceral skeleton*, and the *vertebral column;* and an *appendicular* part consisting of the bones of the limbs, the *girdles*, and the *sternum* or *breastbone* (Holmes, chap. xiii, pp. 229–45).

2. **The vertebral column** (Holmes, pp. 237–38).—The anterior half of the vertebral column of the frog consists of nine bony rings, the *vertebrae;* the posterior half is an elongated slender bone, the *urostyle*. Obtain some isolated vertebrae and identify the following parts:

a) The *neural canal* is the cavity in the vertebra.

b) The *centrum* or *body* is the thickened part of the vertebra ventral to the neural canal. It articulates with the centra before and behind it by a ball and socket arrangement.

c) The *neural arch* is that portion which arches dorsally over the neural canal.

d) The *neural spine* is a sharp dorsal projection from the center of the neural arch.

e) The *transverse processes* are the conspicuous lateral projections extending horizontally outward from the place of junction of centrum and neural arch.

f) The *zygapophyses* are anterior and posterior projections from the neural arch, one in front of and one behind the place of origin of each transverse process. They link successive vertebrae together.

The first vertebra, or *atlas*, differs from the others in the absence of the transverse processes and the presence of special articulating surfaces for holding the skull. Between the atlas and the skull is a gap which permits the operation of pithing. The transverse processes of the last vertebra articulate with the pelvic girdle, and this vertebra is hence called the *sacral* vertebra.

3. **The skull and the visceral skeleton.**—The visceral skeleton includes the upper and lower jaws, the hyoid apparatus, and the cartilages of the larynx. It is so called because originally all of these structures were paired semicircular cartilages used to support the gills. Since gills are part of the walls of the alimentary canal, the appropriateness of the term "visceral skeleton" becomes apparent. Six such cartilaginous hoops are present in the frog tadpole, of which the first becomes in the adult frog the basis of upper and lower jaws, while the remaining ones undergo remarkable transformations into the hyoid and laryngeal cartilages. The laryngeal cartilages will not be studied.

The original cartilage skull, forming a case to inclose the brain, fused in front with the *olfactory capsules* containing the organ of smell, behind with the *otic capsules* containing the ears, and below with the cartilaginous bars which formed the upper jaw while the similar bars of the lower jaw remained separate and formed a joint with the skull. Part of the cartilage of these structures persists in the adult skull, part is converted into cartilage bones, and both of these are partly concealed by a superficial covering of membrane bones.

a) Dorsal aspect of the skull and upper jaw (Holmes, chap. xiii, pp. 229–37).— Examine the dried skull and note that the membrane bones can be distinguished easily from the cartilage bones by their lighter color, smoother surfaces, thin or flat shapes, and more superficial position. The cartilage has of course disappeared in a dried skull. The skull proper forms the central region, while the lower jaw, which is fused to the skull, appears as an arch on each side and in front of the skull. A large gap, the *orbit*, which holds the eye, is thus left between skull and jaw.

The bones of the dorsal side of the skull proper are:

(1) *Nasal* bones, two triangular membrane bones just behind the external nares.

(2) *sphenethmoid* bone, a single ring-shaped cartilage bone in the median line behind the nasals.

(3) *parietal* bones, two long flat membrane bones posterior to the nasals. In other animals the anterior *frontal* portion is separate from the *parietal* portion of this bone.

(4) *occipital* bones, the two cartilage bones forming the posterior extremity of the skull. They surround a large opening, the *foramen magnum*, through

which the medulla oblongata passes to become continuous with the spinal cord. Ventrally they bear two rounded prominences, the *occipital condyles*, by which the skull articulates with the atlas.

(5) *Pro-otic* bones, the cartilage bones extending laterally from the posterior end of the frontoparietals. They are ossified in the otic capsule and hence inclose the ear.

The bones of the upper jaw or *maxillary arch* are:

(1) *Premaxilla*, a pair of small membrane bones in front of the external nares.

(2) *Maxilla*, the long slender membrane bone forming the greater part of the sides of the jaw. Premaxillae and maxillae bear teeth.

(3) *Quadratojugal*, the short slender membrane bone behind the maxilla, not bearing teeth.

(4) *Squamosal*, the curious T-shaped membrane bone extending from the posterior end of the quadratojugal to the pro-otic.

(5) *Pterygoid*, a three-rayed cartilage bone under the squamosal but visible from the dorsal side.

(6) The *quadrate* cartilage lies between the squamosal and the pterygoid and is the place of attachment of the lower jaw to the skull.

b) Ventral aspect of the skull. The ventral bones of the skull are:

(1) *Vomer*, a pair of membrane bones behind the premaxillae, forming the floor of the olfactory capsules and bearing teeth.

(2) *Palatine*, two slender cartilage bones extending laterally from just behind the vomerine teeth to the maxillae.

(3) *Parasphenoid*, a single long dagger-shaped membrane bone on the ventral surface of the skull, its lateral posterior processes underlying the auditory capsules. The point of the dagger underlies the ventral side of the sphenethmoid ring.

c) The bones of the lower jaw or *mandibular arch*. These are:

(1) *Mentomeckelian* bones, two small cartilage bones at the tip of the lower jaw. They are ossified in the original cartilage bars (Meckel's cartilage) which were the lower jaw of the tadpole.

(2) *Dentary*, a short membrane bone behind the preceding on the outer surface of the jaw.

(3) *Angulosplenial*, the long slender membrane bone forming the greater part of the jaw. Its anterior end is under the dentary. Its outer surface behind the dentary is grooved. In this groove is located in life Meckel's cartilage, which articulates with the quadrate cartilages of the skull.

d) Hyoid apparatus. This portion of the skeleton is usually lacking in dried material. Most of its parts have already been seen during the dissection of the frog and should be examined again and further exposed in your preserved specimen (Holmes, p. 324).

(1) The *body* of the hyoid is the flat plate of cartilage in the floor of the hyoid cavity. Its concave anterior margin receives the base of the tongue. Its lateral anterior corners have short *alary* processes, and its posterior corners *postero-lateral* processes.

(2) The *anterior horns* or *cornua* of the hyoid are the long, slender rods which curve back from the anterior margin of the body to the pro-otic bones of the skull.

(3) The *thyroid* processes or *posterior horns* are the two processes which diverge from the posterior margin inclosing the laryngeal chamber between them. They are ossified while the other parts of the hyoid are cartilaginous.

(4) The *columella*, the small slender bone in the middle ear, one end of which is attached near the middle of the tympanic membrane, is probably a part of the hyoid apparatus.

4. Bones of the pectoral girdle (Holmes, p. 238).—The pectoral girdle forms a bony arch for the support of the fore limbs. It is incomplete dorsally and has no connection with the vertebral column. Ventrally its two halves are united by the interposition of the sternum. Each half of the girdle consists of:

a) *Suprascapula*, the most dorsal bone, large and flat with a cartilage along its free dorsal border.

b) *Scapula*, the bone ventral to the preceding and containing a cup-shaped cavity, the *glenoid fossa*, in which the long bone of the upper arm is inserted.

c) *Clavicle*, the anterior bone of the two which compose the ventral aspect of the girdle. It is a membrane bone, the only one in the skeleton outside of the skull and jaws.

d) *Procoracoid*, the cartilage which is covered over by the clavicle and which fails to ossify because its function is usurped by the clavicle.

e) *Coracoid*, the posterior of the two ventral bones. It takes part in the glenoid fossa.

5. The sternum (Holmes, p. 240).—The sternum is a chain of bones and cartilages between the two ventral ends of the halves of the pectoral girdle.

a) *Episternum*, the rounded catilage forming the anterior extremity of the sternum.

b) *Omosternum*, the bone behind the preceding.

c) *Epicoracoids*, the cartilages between the medial ends of the coracoid bones.

d) *Sternum* proper or *mesosternum*, bony rod behind the coracoids.

e) *Xiphisternum*, terminal rounded cartilage.

6. Bones of the pelvic girdle (Holmes, pp. 242–43).—This is the bony girdle which supports the hind limbs. It is complete dorsally, forming a joint with the transverse processes of the last vertebra. The bones of each half of the pelvic girdle are:

a) *Ilium*, the long scythe-shaped dorsal bone which extends forward parallel to the urostyle to articulate with the transverse processes of the sacral vertebra.

b) Pubis, the anterior portion of the semicircular crest which projects ventrally from that part of the pelvic girdle which lies medially between the heads of the two long thigh bones.

c) Ischium, the posterior portion of the crest.

The two pubes and the two ischia are completely fused in the median ventral line producing the projecting crest mentioned above. These unions are named the *pubic* and *ischial symphyses*. For the exact boundaries between the three bones of the girdle see Holmes, Fig. 69, p. 242. The cuplike cavity on each side of the girdle which receives the head of the thigh bone is called the *acetabulum*.

7. Bones of the limbs.—The skeleton of the fore and hind limbs is evidently built upon the same plan, and the bones evidently correspond. The two limbs will therefore be considered together (Holmes, pp. 241, 243, and Fig. 63, p. 230).

a) The upper part of each limb consists of a long bone. In the fore limb this is the *humerus;* in the hind limb, the *femur*. The humerus bears a conspicuous crest, the *deltoid ridge*, so named because the deltoid muscle is inserted there.

b) The next section of the limb is generally composed of two bones, but in the frog these two are in the case of both limbs fused into one. This is the *radio-ulna* in the fore limb, *tibio-fibula* in the hind limb. A longitudinal groove along the center of both surfaces of each of these bones indicates the place of fusion of the two originally separate components, the *radius* and *ulna* in the forearm, the *tibia* and *fibula* in the shank. Radius and tibia correspond and are on the thumb side (*preaxial* side) of the limb; ulna and fibula are on the little finger side (*postaxial*).

c) Wrist and ankle constitute the next section of the limbs. The wrist or *carpus* consists of six small bones in two rows. These are generally difficult to make out in ordinary preparations of the skeleton. The ankle or *tarsus* is unusually elongated in the frog and consists chiefly of two relatively large bones, the *astragalus* on the preaxial side and the *calcaneum* on the postaxial side. Between these and the foot, two or three minute bones occur.

d) The palm of the hand and sole of the foot each consist of five slender diverging bones, the *metacarpals* and *metatarsals*, respectively. The first metacarpal is rudimentary.

e) The fingers and toes are supported by small bones, called *phalanges*.

I. THE MUSCULAR SYSTEM

The frog has two types of muscles, the *involuntary* muscles found in the viscera, and the *voluntary* ones attached either directly or indirectly to the skeleton. The arrangement of the muscles of the viscera has already been seen in the study of the microscopic structure of the organs. They are commonly arranged in cylindrical tubes, in which the fibers run in either a circular or a longitudinal direction. The voluntary muscles, on the other hand, have no

such simple arrangement, and a study of each one is necessary to understand
its action. The term "muscular system" generally refers to the voluntary
muscles.

For the study of the muscles frogs which have been preserved for some
time in formalin should be employed. Those in which the viscera have been
dissected are usable for this purpose. Remove the skin completely.

1. **Parts and relations of a muscle.**—Examine the large *gastrocnemius* muscle
on the back of the shank as an illustration of a typical muscle. Identify its
parts as follows:

a) The *fascia* is the shining tough connective tissue membrane which incloses
the muscle.

b) The *tendon* is the shining tough band or cord at each end of the muscle.
It is produced by the extension and concentration of the fascia beyond the
fleshy part. By means of tendons, muscles are firmly attached to bones, other
tendons, or other fixed structures. Tendons may be of many shapes, depending
upon the shape of the muscle of which they are a part.

c) The *belly* is the fleshy part of the muscle.

d) The *origin* is the more fixed point of attachment of the muscle, i.e., the
part which does not move when the muscle contracts. The gastrocnemius has
two points of origin, or *heads*, as they are often called, at the upper end of its
(apparently) dorsal surface. The larger of these attaches the main mass of the
muscle to a tendon which passes from the distal end of the femur to the upper
end of the tibio-fibula; the smaller is a slender tendon which joins the general
tendon passing over the knee. With a forceps loosen the other muscles from
about the heads of the gastrocnemius and verify these points.

e) The *insertion* is the more movable point of attachment of the muscle,
i.e., the part which is moved by the action of the muscle. The lower end of
the gastrocnemius tapers to a strong tendon (the famous *tendon of Achilles*),
which passes over and is attached to the ankle bones and then becomes continuous
with the broad *plantar* fascia which covers the sole of the foot.

f) The *action* of a muscle is a description of its function. The gastrocnemius
through its insertion on the ankle can bend the entire leg below the knee up
against the thigh (*flexion* of the leg); through its continuity with the plantar
fascia it can straighten the foot (*extension* of the foot). Test these actions by
pulling on the gastrocnemius.

2. **General account of the muscles of the frog** (Holmes, chap. xiv, pp. 246–
57).—We shall undertake to study only the superficial and easily identifiable
muscles, particularly those which are of interest in the physiology of the frog.
Most of the following muscles have the same names, positions, and action as
in man and other vertebrates, and the statements about them are hence of general
application. No attempt is made to give all the details of the origin and insertion.

In studying muscles separate each one from its fellows as carefully as possible with a forceps or probe, and determine the origin, insertion, and general action of each.

a) Muscles of the lower jaw: Pull off all tissue between the eye and tympanic membrane. Pull off the tympanic membrane, identifying underneath it a circular cartilage, the *tympanic ring,* resting upon the squamosal bone. Between the tympanic ring and the eye a mass of muscles will be found passing to the lower jaw, on the inner side of the posterior end of the upper jaw. Remove the end of the upper jaw (quadratojugal bone) so as to reveal the complete course of these muscles. The following three may be readily identified:

(1) The *temporal* muscle arises from the side of the skull and passes down between the eye and the tympanic ring to be inserted on the posterior end of the lower jaw. Action, closes the mouth (*elevator* of the jaw).

(2) The *masseters* originate from the tympanic ring and adjacent bones and are inserted on the lower jaw behind the temporal. Action, same as preceding; also stretch the tympanic membrane.

(3) The *depressor mandibuli* is a muscle of the jaw located behind the tympanic ring. It arises from the tympanic ring and from the general fascia of the back (*dorsal fascia*) and passes to the extreme posterior tip of the lower jaw, where it is fastened to Meckel's cartilage. Action, opens the mouth (*depressor* of the jaw) and stretches the tympanic membrane. Pull the lower jaw fully open to see more clearly the arrangement of these three muscles.

b) Muscles of the dorsal side of the trunk: This portion of the body is more or less covered by the dorsal fascia, a strong membrane fastened to the ilium bones, the neural arches of the vertebrae, and the skull, and furnishing a place of insertion of many muscles. Easily identifiable muscles of the back beginning behind the ear are:

(1) The *dorsalis scapulae* is the anterior half of the large triangular mass behind and partially covered by the depressor mandibuli. Origin, from the dorsal margin of the suprascapula; remaining course like the next.

(2) The *latissimus dorsi* is the posterior portion of the triangular mass. Origin, dorsal fascia; unites with the preceding to be inserted on the deltoid ridge of the humerus; action, raises the fore limb upward and backward (abduction of the limb).

(3) The *longissimus dorsi* is the long muscle extending from the anterior third of the urostyle forward to the skull. Remove the dorsal fascia and preceding muscles to see its full course. It is attached at many places to the vertebrae. Action, raises the head and straightens the back.

(4) The *coccygeo-sacralis* runs diagonally from the urostyle just behind the insertion of the preceding muscle to the transverse process of the sacral vertebra. Pull off the remainder of the dorsal fascia to see it. Action, draws the back

the urostyle or vice versa when both muscles of the two sides act together, or turns the back to one side, when acting singly.

(5) The *coccygeo-iliacus* is the diagonal muscle parallel to and behind the preceding running from the posterior two-thirds of the urostyle to the ilium. Action, fixes the urostyle with respect to the pelvic girdle.

c) Muscles of the sides of the abdomen:

(1) The *external oblique* is the large muscle covering the sides of the abdomen extending from the dorsal fascia and the ilium to the linea alba. Its fibers run diagonally backward. Action, to support and reduce the abdominal cavity, and to cause exhalation in lung breathing.

(2) The *transverse* (including the *internal oblique* of other vertebrates) lies under the preceding, its fibers running diagonally forward. Strip off the external oblique cautiously to see it. Relations and function similar to preceding.

d) Muscles of the ventral side of the trunk:

(1) The *rectus abdominis* is the flat segmented muscle lying on both sides of the median ventral line, extending from the pubic symphysis to the sternum. Action, supports the abdominal contents, and fixes the sternum in place.

(2) The *pectoral* muscle is the large muscle of the anterior ventral part of the body. It arises from the various parts of the sternum and from the lateral border of the fascia of the rectus abdominis muscle. It is inserted on the deltoid ridge of the humerus. Action, draws the arm toward the ventral side and leg (adductor of the arm) and expands the abdominal cavity.

e) Muscles of the floor of the mouth and the hyoid apparatus:

(1) The *mylohyoid* is a thin sheet of muscle running crosswise from one half of the lower jaw to the other along their whole extent. It forms the outermost layer of the buccal floor and its function is to raise this floor in the breathing movements.

(2) The *submental* muscle. Cut through the mylohyoid muscle in its median ventral line and deflect the two halves. Under it at the very anterior tip of the lower jaw is the small submental muscle. Its contraction pushes the sublingual tubercle upward against the premaxillary bones of the upper jaw and thereby closes the external nares in lung respiration.

(3) The *geniohyoid* muscle comprises the longitudinal bands revealed by the removal of the mylohyoid, consisting of medial and lateral portions arising under the submental muscle and lateral to it from the lower jaw, and extending to the postero-lateral processes of hyoid and the body of the hyoid as far as the forking of the thyroid processes. Action, pulls the hyoid apparatus powerfully forward, thus raising the floor of the buccal cavity in respiration; also helps to swallow, to open the mouth, to lower the tip of the jaw, thus opening the nares, and to move the tongue.

(4) The *sternohyoid* is a continuation forward of the rectus abdominis, extending from the underside of the coracoid and clavicle to the body of the

hyoid. It is readily seen when the pectoral girdle is lifted up. It and a small muscle lateral to it (*omohyoid*) exert a pull upon the body of the hyoid, causing it to bulge outward, and hence lower the floor of the buccal cavity in breathing. Their action is thus the opposite of that of the preceding muscles.

(5) The *petrohyoids* are several small muscles under the sterno- and omohyoids extending from the otic capsule to the sides of the hyoid apparatus. They raise the hyoid apparatus and hence the floor of the buccal cavity in respiration, acting in antagonism to the preceding muscle, and by their compressing effect upon the larynx and pharynx are of great importance in swallowing food and air.

f) Muscles of the tongue:

(1) The *hyoglossus* is the conspicuous muscle in the median ventral line of the throat under the geniohyoid. Each half of it originates at the posterior end of the thyroid process of the hyoid, extends forward covering this process, meets its fellow where the processes spring from the body of the hyoid. The muscle thus formed runs forward in contact with the body of the hyoid up to the base of the tongue into which it disappears. It is the *retractor* of the tongue (draws it back into the mouth after use).

(2) The *genioglossus* is a small but thick muscle lying in front of the anterior end of the hyoglossus and originating from the lower jaw under the submental muscle. It is the *protractor* of the tongue (throws it forward).

g) Muscles of the thigh: The thigh presents apparent dorsal and ventral sides and anterior and posterior surfaces. These are not really such because the leg of the frog has been twisted from the primitive vertebrate position. The ventral surface is really inner and hence more correctly called *preaxial;* and the dorsal surface is outer or *postaxial.* Anterior is really ventral and posterior dorsal. However, in order to simplify the following description, the same names will be applied as for other parts of the body, according to the apparent positions. Separate all of the muscles of the thigh from each other before proceeding.

(1) The *triceps femoris* is the great muscle which covers the whole anterior part of the thigh, its powerful tendon passing over the knee to the tibio-fibula. It has three origins or heads and consists of three parts. The ventral part (*vastus internus* or *crural*) arises from the borders of the acetabulum. The small middle portion (*tensor fasciae latae*) originates on the ilium and ends in the fascia (*fascia lata*), which covers the triceps femoris. The dorsal part is the *vastus externus* or *gluteus magnus* and arises from the side of the posterior end of the crest of the ilium. The triceps femoris is the great extensor of the shank, and also may draw the leg up against the body (*abduction*).

(2) The *iliacus* muscles are those which extend from a considerable part of the crest of the ilium between the tensor fasciae latae and the vastus externus to the femur. They are abductors of the leg.

(3) The *ileo-fibularis* is the slender muscle next to the vastus externus on the dorsal side of the thigh. It extends from the ilium to the upper end of the tibio-fibula, draws the thigh dorsally, and flexes the shank.

(4) The *semimembranosus* is the posterior muscle of the dorsal side of the thigh. It arises from the ischial symphysis, is inserted on the tibio-fibula, bends the shank, and draws the leg toward the median line (*adduction*).

(5) The *gracilis major* and *minor* (*rectus internus major* and *minor* in Holmes, Fig. 70, p. 249) are the posterior muscles of the ventral side of the thigh on the other side of the leg from the preceding. The minor is small and the most posterior one. Both extend from the ischium to the knee and have the same action as the preceding muscle.

(6) The *adductor magnus* is the muscle next anterior to the gracilis major on the ventral side. Most of it is concealed by gracilis major and the muscle to be mentioned next. It originates on the ischial and pubic symphyses and is inserted on the femur. It adducts the thigh and leg.

(7) The *sartorius* is the flat thin muscle crossing the lower end of the adductor magnus. It arises on the pubic symphysis and joins the general tendon over the knee. Action, bends the shank and aducts the thigh.

(8) The *adductor longus* is a thin flat muscle under the sartorius but generally peeping out along the latter's anterior border. It originates on the ilium, joins and acts with the adductor magnus. .

(9) The *semitendinosus* is a muscle of peculiar shape under the gracilis major, which should be removed to see it. It has two separate tendinous heads from the ischium and two separate bellies, uniting to one tendon fastened to the upper end of the tibio-fibula. It acts like the gracilis major.

h) Muscles of the shank:

(1) The *gastrocnemius* has been sufficiently described.

(2) The *peroneus* is the only other muscle on the dorsal aspect of the shank. It extends from the general tendon over the knee to the lower end of the tibio-fibula and the ankle. It extends and twists the foot and brings the shank up against the thigh, as in swimming and leaping movements.

(3) The *tibialis* muscles are the small muscles of the shank lying next to the bone. There are three of them. The *tibialis anticus longus* covers the anterior surface of the tibio-fibula; it arises by a slender tendon from the lower end of the femur and soon divides into two bellies, which are attached by slender tendons to the ankle bones. Action, bends the ankle. Beneath the lower part of this muscle is the small *tibialis anticus brevis*, originating on the middle of the tibio-fibula, and inserted on the ankle, which it also flexes. The *tibialis posticus* is a long slim muscle on the ventral aspect of the shank, between the gastrocnemius and the tibio-fibula, to which it is attached along its entire length. It is likewise inserted on the ankle, which it flexes and twists.

(4) The *extensor cruris* is a small muscle on the anterior aspect of the shank, lying next to the upper two-thirds of the tibio-fibula and in contact dorsally

with the tibialis anticus longus. It originates on the femur and straightens the shank in the leaping and swimming movements.

J. GENERAL ANATOMICAL PRINCIPLES

Now that we have completed our study of the anatomy of the frog, attention may be called to some of the general principles which underlie its construction.

1. Principle of bilateral symmetry.—The parts of the frog are arranged symmetrically with reference to a median vertical plane which was named in the early part of this outline the sagittal plane. These parts are either single (*unpaired*) and lie in the sagittal plane which divides them into identical right and left halves, or they are double (*paired*) and placed at the same level of the body at equal distances from the sagittal plane. Unpaired structures are the skull and vertebral column, brain and spinal cord, heart, digestive tract, postcaval vein, and dorsal aorta. All the muscles, the appendicular skeleton, the nerves and chief sense organs, most of the blood vessels, the lungs, kidneys, reproductive organs and their ducts are paired. The digestive tract is the chief unsymmetrical system in the body, but it obviously began as a median tube extending from mouth to anus and only subsequently developed the lateral displacements and spiral coilings which have destroyed its symmetry.

2. Principle of segmentation.—Less readily recognizable is the fact that the structure of the frog is based upon a repetition of parts along the sagittal axis. The frog is thus conceived of as built up of a series of sections, or *segments*. These segments are similar to each other, each has perfect bilateral symmetry and contains a portion of each of the systems of the body. In the adult frog segmentation is best illustrated by the spinal cord and its nerves, the vertebral column, and some parts of the circulatory system (vessels to the body wall, kidneys, and reproductive organs), all of which exhibit obvious repetition along the axis, and to a less extent in the muscles (rectus abdominis and longissimus dorsi muscles). The segmentation is much more complete in the tadpole.

3. Principle of cephalization.—Segmentation is retained most completely in the posterior portion of the body, less so in anterior regions, and is almost entirely lost in the head. Investigation shows, however, that the head like the rest of the body originally consisted of a series of segments, and traces of this segmentation still persist in the lobed condition of the brain, in the cranial nerves, and muscles of the eye. But these segments for the most part fused together in order to produce a structure, the head, which should be more specialized, more efficient than the other parts over which it acquires dominance, just as men and nations combine together for greater efficiency and achievements. Correlated with this dominance of the head and anterior regions is a descent of the viscera posteriorly. *Cephalization* is the name which is applied to the development and specialization of the head at the expense of the rest of the body.

VI. THE PROCESS OF CELL DIVISION

All cells arise from pre-existing cells by a process of division. In a few cases this is a simple splitting, or *direct division*, but in the vast majority of cases the cell divides by a complicated process, known as *indirect division, mitosis*, or *karyokinesis*. In mitosis, both the nucleus and the cytoplasm are involved in a complex and remarkable behavior.

The following materials have been found to be the most favorable for the study of mitosis: the developing eggs of *Ascaris*, a parasitic worm; the root tips of plants, chiefly the onion, the hyacinth, or *Tradescantia;* the developing eggs of fish. A complete outline is provided below for the study of mitosis in the eggs of *Ascaris;* and following this are brief statements regarding the other materials and the points in which they differ from *Ascaris*.

A. MITOSIS IN THE EGGS OF *Ascaris*

Ascaris megalocephala, commonly used for this purpose, is a parasitic round worm found in the intestine of the horse. (A description of this animal is given in Hegner, p. 160.) The fertilized eggs pass down the long oviducts of the worm, dividing as they go. Obviously by cutting longitudinal slices through the oviducts at the proper levels, eggs in all stages of division will be obtained. The slides bear such longitudinal slices of the oviduct.

Examine the slide "*Ascaris*—mitosis" with the low power. Identify in each long slice upon the slide the thin walls of the oviduct, composed of large epithelial cells, and its wide cavity completely filled with round objects, each of which is an egg inclosed in a thick shell. Examine one of the round objects with the high power and get a clear idea of what you are looking at. Identify in each one the thick *shell* inclosing a cavity in which floats the *egg cell*, considerably smaller than the cavity. The egg has been fertilized and hence possesses two nuclei, its own nucleus and the nucleus from the sperm. Its cytoplasm is *vacuolated*, that is, appears to contain a number of empty spaces. Examine the egg cells with your highest power and look for each of the following stages of mitosis (see Hegner, p. 29). Considerable searching may be required to find the various stages. Have the assistant help you. As the sections are very thin, only one nucleus may appear, or only parts of the mitotic apparatus may be present. Avoid drawing such partial pictures. Make your drawings large and detailed.

1. **The resting cell.**—In the so-called *resting state*, that is, the condition before mitosis begins, the egg presents the same appearance as other cells which we have studied. It contains two nuclei, its own and the sperm nucleus. Each

of these is a rounded vesicle, inclosing the usual chromatin granules and nuclear sap. **Draw** such a stage in detail. The egg shell may be omitted.

2. Spireme stage or early prophase.—As cell division begins the chromatin granules thicken and mass together, finally uniting into a long, coiled thread, called the *spireme*, which fills the nucleus. Consider that in sections only pieces of such a coiled thread could appear. Look, therefore, for nuclei containing deeply stained elongated pieces of chromatin. At this time also there appear in the cytoplasm two dense collections of granules, called the *asters*. Each aster has in its center a black granule, the *centrosome*, and sends out radiations, the *astral rays*, into the surrounding cytoplasm. A complete picture of the spireme stage shows the egg and sperm nuclei, containing broken threads of chromatin, in contact with each other, and an aster placed at each end of the plane of contact. You may not be able to find a cell cut in the right plane to show all of these parts simultaneously. **Draw** in detail the best example you can find.

3. Late prophase.—The spireme thread now breaks into a number of distinct threadlike bodies, each of which is called a *chromosome*. In *Ascaris* there are four of these. The nuclear membrane has disappeared and the chromosomes lie free in the cytoplasm. They stain very deeply, and are usually U-shaped. Meantime the asters have drawn farther apart and delicate fibrils extend between them. These fibers are called the *spindle*, and the whole structure, asters and spindle, is known as the *mitotic figure*. **Draw** a cell showing the four chromosomes free in the cytoplasm.

4. Metaphase.—The chromosomes now arrange themselves in the center of spindle. The mitotic figure is now fully developed and symmetrically placed in the cell. Find a cell which is cut parallel to the spindle and **draw**, showing spindle, asters, and the band of chromosomes across the center of the spindle. Each chromosome has a longitudinal split at this time, which is generally difficult to see.

5. Anaphase.—Each chromosome next splits in two longitudinally and the two halves separate. One-half of each chromosome moves toward one aster and the other half to the other aster. In this migration the ends of the chromosomes always point toward the middle of the spindle, so that the dividing cell in this stage contains two groups of chromosomes, each group looking something like the top of a palm tree, with the delicate parallel threads of the spindle stretching between them. **Draw.**

6. Telophase.—The chromosomes approach the asters, where each group condenses into a mass in which the individual chromosomes are no longer distinguishable. A constriction which gradually pinches the cell into two equal parts is appearing midway between the two chromatin masses. **Draw.**

7. Completion of mitosis.—The constriction deepens dividing the cell into two cells, the chromatin mass resolves itself into the ordinary nuclear structure, and a nuclear membrane is formed. Two cells each exactly like the original single cell have thus been produced.

B. MITOSIS IN PLANT ROOT TIPS

The root tips of plants grow very rapidly in some cases and hence are favorable places for finding all the stages of mitosis. For this purpose longitudinal sections are made through the root tip. The tip is covered by a *root cap* consisting of older and hardened cells, which is pushed ahead of the real growing region. The dividing cells are therefore found a little distance back from the tip.

The process of mitosis in these plant cells differs from that seen in the eggs of *Ascaris* chiefly in that centrosomes and asters are entirely lacking, so that the mitotic figure consists of spindle only. Further differences are: in the resting stages one or more conspicuous nucleoli will be found within the nucleus; the spireme thread is closely coiled so that at this stage the section of the nucleus is packed with small wormlike segments of the spireme; the chromosomes are numerous and less distinctly U-shaped; in the anaphase the resemblance of each group to the tops of palm trees is quite striking; and in the telophase, the new cell wall is not produced by a constriction, but forms in place, apparently in part from a condensation of the spindle fibers. The spindle is rather faint throughout and is seen best only in the late stages of mitosis.

C. MITOSIS IN THE EGGS OF THE WHITEFISH

The sections are taken through the dividing eggs of the fish, generally in the early stages when the cells are quite large. Each section shows a number of cells, some of which will be found to be in various stages of mitosis. Differences from *Ascaris* are that the chromosomes are quite small, very numerous, and rodlike in form; and that a number of minute centrosomes instead of the one large centrosome of *Ascaris* are present. They are difficult to see. A clear field surrounds the place occupied by the centrosomes, and from this the long fibers of the asters radiate nearly filling the cell. Fish eggs are valuable for the study of mitosis chiefly because the mitotic figure is of such large size and so distinctly fibrillar in them that they present a striking appearance, which closely corresponds to the textbook representations of this structure.

VII. GENERAL EMBRYOLOGY

The egg cell, after the sperm cell has penetrated it (process of *fertilization*) enters on a process of *development*, the study of which constitutes the subject of *embryology*. An elementary study of embryonic development will be made first on a simple case like that of the starfish (*Asterias*), then on the more complicated case of the frog (Hegner, pp. 107-111).

A. DEVELOPMENT OF THE STARFISH

The fertilized egg first divides by mitosis a number of successive times until a large number of very small cells is produced. This part of development is called *cleavage* or *segmentation*. Study slide "*Asterias*—early cleavage." Note that the objects on the slide are *not* sections but the entire cells. (The cells which exhibit a large clear nucleus containing a black spot [nucleolus] are unfertilized eggs). Each egg and embryo is surrounded by a membrane, which is called the *fertilization membrane*, and is separated from the egg at the time of fertilization.

1. **Two-celled stage.**—The egg divides into two equal halves, which remain in contact with each other. **Find and draw in outline.**

2. **Four-celled stage.**—A second division occurs at right angles but passing through the same axis as the first. **Draw.**

3. **Eight-celled stage.**—Each cell of the four-celled stage divides in two transversely at right angles to both the previous divisions, producing eight equal cells in two plates of four cells each. **Find and draw.** Note that the cells may easily become displaced from their natural position in the making of the slide, and pick out only those that present the normal appearance.

4. **Later cleavage.**—Examine slide "*Asterias*—late cleavage." The process of division continues until a large number of cells is produced. Meantime, a central cavity appears between the cells.

5. **Blastula.**—Examine slide "*Asterias*—blastula or larval stages." At the end of the cleavage process, the embryo consists of a single layer of cells surrounding a central large cavity, the *segmentation cavity*. This stage is the *blastula*. Its form is that of a rubber ball. Find one of these balls of small cells, and focus so as to obtain an *optical section*, i.e., focus so that the appearance is the same as if you had actually made a section through the center of the blastula. In such a focus the blastula appears as a circular layer of cells, the layer being one cell thick, surrounding a large cavity. **Draw the optical section.**

6. Gastrula.—Examine slide "*Asterias*—gastula or larval stages." This stage marks the end of the cleavage stage (although of course cell division continues) and the beginning of *differentiation*. The *gastrula* arises from the blastula by the pushing in of one wall just as if one should thrust one's finger into the side of a rubber ball. The gastrula appears on the slide as an oval body, containing a central darker projection. Find one in profile, obtain an optical focus on it and **draw**. The outer layer of cells is called the *ectoderm;* the inner layer, which has been pushed in or invaginated, the *entoderm;* the entire invaginated structure is the *archenteron*, or primitive intestine; and the opening of the archenteron to the outside is the *blastopore*. The next step in development is the formation of a third layer of cells, the *mesoderm*, between the ectoderm and entoderm, and these three layers are called *germ layers* because from them all the structures of the adult organism are derived.

B. DEVELOPMENT OF THE FROG

Preserved material will be provided for this purpose. Remove the jelly from the eggs. Consult Holmes, chapter v, pp. 89–103.

1. Cleavage stages.—The egg of the frog is much larger than that of the starfish because it contains a considerable quantity of yolk, a semifluid food material containing proteins and fats. The egg is black and white, the black half constituting the *animal hemisphere* and consisting largely of protoplasm, while the white half, the *vegetative hemisphere*, holds most of the yolk. Owing to the presence of yolk, the vegetative hemisphere divides more slowly and produces larger cells than the animal hemisphere.

Study with the low power or with a hand lens and **draw** two-, four-, and eight-cell stages. Note the inequality in the size of cells of the two hemispheres, and the tendency of the vegetative cells to lag behind, so that six cell stages may occur.

2. Blastula.—After numerous cleavages a blastula is produced consisting of numerous very small black cells, and less numerous larger white cells. **Draw.** Then with a sharp knife bisect the blastula by a cut which cleaves the black and white hemispheres, and examine the cut surface with a hand lens. Compare with Holmes, Fig. 18 (p. 93), and **draw**. The blastula of the frog differs from that of the starfish in that its wall is composed of several layers of cells.

3. Gastrula or yolk plug stage.—Owing to the large amount of inert yolk in the white cells, they cannot invaginate as in the case of the starfish, but the gastrula is formed mainly by the growth and extension of a sheet of black cells down over the white cells. The white cells are thus inclosed and become the entoderm. The closure of the black cells over the white is not quite complete, leaving a circular opening, the blastopore, through which some of the white cells protrude, producing a circular white area, the *yolk plug*. **Draw,** showing the blastopore and yolk plug. Bisect the gastrula by a cut through the yolk plug and the center of the black hemisphere. Examine the cut surface, compare with Holmes's Figs. 19 and 20 (pp. 94 and 96), and draw. The black cells are the ectoderm, the white cells, the entoderm, and about this time a third layer, the mesoderm, begins to grow out between the ectoderm and entoderm from the cells around the blastopore.

4. **Origin of the nervous system; neural fold stage.**—Examine embryos of this stage with a lens and note that a fold is appearing on each side of a central groove extending lengthwise along the black hemisphere. This groove marks the dorsal median line of the future embryo. The pair of folds, the *medullary* folds, later come in contact and fuse in the median dorsal line, thus forming a longitudinal tube which is the central nervous system. The blastopore is now reduced to a small hole. **Draw.** Holmes's Fig. 22 (p. 98) is a cross-section of this stage.

5. **Mesoderm and coelome.**—Examine slide. This is a cross-section of a stage about halfway between Holmes's Figs. 22 and 26. The cross-section is roughly pear-shaped, the narrow end of the section being the dorsal side of the embryo. The section is surrounded by a layer of uniform width, two or three cells thick. This is the ectoderm, destined to become the epidermis of the skin. In the dorsal median line just under the ectoderm is an oval mass of cells with a central elongated cavity. This is the cross-section of the neural tube or central nervous system, which, as has been seen, originates as a pair of ectodermal folds, which then unite to form a tube, while the ectoderm fuses again over the tube to a continuous layer. Just ventral to the tube is a circular mass, the *notochord*, which arises by an upfolding of the dorsal wall of the intestine. It is a long, slender rod, around which the vertebrae later develop. Ventral to the notochord appears the primitive intestine, a large, nearly circular, mass of ill-defined cells. Its dorsal wall is thin and overlies a cavity, the cavity of the future digestive tract; its ventral wall is very thick on account of the yolk which it contains. This intestine is the entoderm, and its cells become the lining epithelium of the digestive tract. Between the ectoderm and the entoderm lies the mesoderm; it consists of a large triangular mass of cells on each side of the neural tube from which a layer of cells extends ventrally on either side of the intestine, meeting below. The layer of mesoderm cells generally shows a central split, which is the coelome. The mesoderm to the outer side of the split, next to the ectoderm, then represents the parietal layer of the peritoneum and some of the connective tissue of the body wall. The mesoderm on the inner side of the split, next to the entoderm is destined to form the connective tissue and muscular layers of the wall of the digestive tract, the visceral layer of peritoneum, and the mesenteries. The masses of mesoderm at the sides of the neural tube and notochord form the axial skeleton, nearly all of the voluntary muscles, and the dermis of the skin. The mesoderm immediately ventral to the triangular mass is the source of the urinogenital system.

Draw the cross-section. Do not attempt to put in individual cells, as these are not distinctly visible.

6. Later development and segmentation.—Examine with the hand lens embryos of about the stages of Holmes's Fig. 24 (p. 101) and Fig. 30, 3 (p. 117). Note the elongation of the body with later appearance of a tail. In the younger embryo observe that the nervous system is sharply marked off along the whole dorsal side and that grooves, later to break through as *gill slits*, are present on the sides of the head. In the older embryo identify the eyes on the sides of the head, the ventral mouth with horny lips, the two suckers posterior to the mouth, and the much-coiled intestine visible through the skin. Note especially in the tail the zigzag *segments*. Each such segment is a unit of structure and will give rise to a vertebra, a section of the spinal cord with a pair of spinal nerves, a certain number of muscles, paired branches of the chief blood vessels, etc. These segments are also present in the body region of the tadpole, although externally invisible. Segmentation, that is, repetition of structures along the axis, in fact underlies the whole make-up of the adult frog.

VIII. HEREDITY: MENDEL'S LAW

In the experiment on the life-cycle of the fruit fly (*Drosophila*) you were given a pair of flies which differed from each other in a single character, as long wings and short wings. (If blowflies were provided for that experiment, a pair of fruit flies will now be given to you and should be examined according to the directions under II, F, 1.) The offspring resulting from such a pairing between unlike individuals are therefore *hybrids*, and it becomes a matter of great interest to find out what will be the appearance of the offspring, as we may then discover how the characters of animals behave in heredity.

A. FIRST HYBRID GENERATION

When the offspring appear note carefully the character of the wings (or other feature which was selected for the experiment). Are they all alike, or of two kinds, like the parents, or do they resemble one parent more than the other, or are they intermediate? What is the meaning of the terms "dominant" and "recessive" as applied to a pair of characters such as we are dealing with here (A)?

A fresh bottle with banana will be provided for each student into which he is to put one or two pairs of flies from his first generation and raise a second generation. The transference of the flies may be accomplished by dropping a small bit of cotton soaked in ether into the bottle and taking out the flies after they have become unconscious, or by holding a bottle over the end of the old culture bottle until a few flies have flown into it, then holding the bottle over the new culture bottle until they have flown out again. In doing this make use of the tendency of *Drosophila* to fly toward the light.

B. SECOND HYBRID GENERATION

In the second hybrid generation note again the character of the wings (or other feature selected) and determine the number of individuals with each kind of wing length. This behavior of a hereditary character is called *Mendel's law* or *alternative inheritance*. If we suppose that every female fly of the first hybrid generation gave rise to two kinds of eggs in equal numbers, one bearing the character "long wings" and the other the character "short wings"; and that similarly every male of the first hybrid generation produced two kinds of sperms in equal numbers; and that in fertilization all possible chance combinations occurred (i.e., each kind of sperm fertilizes both kinds of eggs), what kind and proportions of offspring would we expect mathematically? Thus if L is taken to represent the character "long wings" and s the character "short wings" we get as a result of algebraic multiplication the following possible combinations in equal numbers: LL, Ls, Ls, ss. What will be the external appearance of the flies containing each of these combinations? Will LL differ from Ls? Why? What then will be the ratio of long- to short-winged individuals in the second generation (Hegner, p. 289)?

These experiments were first performed by a monk named Mendel, who made crosses between different varieties of peas in the garden of the monastery at Brünn, Austria. He discovered that in such crosses the differences between the two plants that are crossed do not become permanently blended in the offspring but separate out again unchanged in the second generation. This behavior of the characters in crossing two different organisms is called Mendel's law, and Mendel's results have since been confirmed on a very large number of plants and animals. The conclusion which has been drawn from these experiments is that the characteristics of animals are separate; that they are, so to speak, independent units, which exist in some fashion or other as units in the eggs and sperms; that these units remain separate in a hybrid organism even though the hybrid may appear externally to be a blend of the characters of its two unlike parents; and that under proper circumstances they may be made to separate out again apparently unchanged. Modern research indicates that the unit characters or, more correctly speaking, materials which represent the unit characters are located in the chromatin of the nucleus.

IX. PHYLUM PROTOZOA

A. INTRODUCTORY REMARKS

In the preceding sections of this manual we have studied in detail the anatomy of a fairly complex animal, the frog. We have seen the systems of organs of which the frog is composed, the cells and tissues which are the framework of these organs, and for what purpose and in what manner these organs are used in enabling the animal to continue its existence. We have further seen that all of this complicated mechanism arises from a single undifferentiated cell, the egg, which, stimulated to activity by the entrance into its substance of a sperm, starts on a process of development in the course of which the multitude of structures found in the adult animal come into existence.

In this course of development in animals we further note that certain fundamental steps are involved. First the egg proceeds to produce a large number of apparently similar undifferentiated cells by the process of cell division, which we also studied. This continues until a ball of cells is produced. Then occurs the first step in differentiation; part of the ball invaginates so that a layer of cells, now called the entoderm, lies within another layer of cells, the ectoderm. So important are these layers for the future development that they are designated as *germ layers*, i.e., layers from which certain systems are to arise. Owing to their different positions these two layers have different relations to the external environment and hence must take on different functions. The ectoderm, being in contact with the environment, must necessarily receive stimuli from this environment and act as protection against the harmful conditions which may arise; hence it is destined for nervous and covering structures. The entoderm naturally takes on digestive functions, since food is essential to life, and an animal can hardly digest food unless it takes the food into its interior. This structural condition found in the gastrula stage of development is known as *diploblastic* (meaning two germ layers), and, as we shall see, thousands of animals exist whose structure has gone no farther than this. However, it is obvious that no very great degree of complexity and differentiation can be attained in a gastrula; a third germ layer, the mesoderm, next arises between the other two, from which is produced by far the greater part of the structures that we have seen in the frog. This condition is known as *triploblastic* (meaning three germ layers). The next advance is the splitting of the mesoderm, so as to leave a cavity, the coelome, between its two layers; and finally the mesoderm becomes segmented, that is, repeats itself along the axis.

Having thus established in our minds a fairly complete picture of the make-up of an animal and the manner in which its anatomical features have come into

71

existence, we shall next study a number of other common animals, comparing them with the frog in structure, histology, and function. Just as the frog forms one of a great group of animals, the *vertebrates*, distinguished by the posses- sion of a vertebral column, so all the animals which exist upon the earth can be arranged into great groups, each of which is called a *phylum*. There are about a dozen of these phyla and they are distinguished from each other very easily by large important anatomical differences. We shall study representatives of several of these phyla, starting with the simplest, always comparing their anatomy with that of the frog and seeing how one by one the systems of which the frog is composed have come into existence.

We trust the student has by this time been sufficiently impressed with the fact that the unit of structure of the frog is the cell and that the entire frog comes from one cell. We are now about to see that numerous adult animals exist which consist of but a single cell, but yet are able to carry on all of the functions necessary to life within the limits of this single cell. These animals belong to the phylum *Protozoa*, or the one-celled animals. They are the simplest animals, and one of the simplest among them is the one to which we shall first direct our attention, the *Amoeba*. Read Hegner, chapter iv, pp. 37–53.

B. THE AMOEBA

1. **General structure and locomotion.**—Mount a few drops of solid material from the culture on a slide, cover, and examine with the low power. *Cut down the light.*   Search the slide for an irregular granular object, apparently motionless. Ask the assistant whether or not you have an *Amoeba* or have him help you find one. Study the animal with both low and high powers. Observe that the *Amoeba* moves by putting out projections, called *pseudopodia*, from its surface and then flowing into these projections. This type of movement is designated as *amoeboid movement*. The protoplasm of the *Amoeba* may be divided into two regions, an outer clear layer, the *ectoplasm*, entirely free from granules, and the central mass, the *endoplasm* (also spelled "entoplasm"), filled with round or dark oval granules, of unknown function, and food particles. Under the high power watch the formation of a pseudopodium, and determine how the ectoplasm and endoplasm behave in its formation. Does the Amoeba have anterior and posterior ends or a definite form?   •

Make five outline drawings of the Amoeba to show successive changes of shape. Indicate by arrows the direction of flow of the protoplasm.

2. **Special structures.**—Find a specimen whose protoplasm is well spread out and not excessively granular and look with the high power for the following structures:

a) *The contractile vacuole:* Watch the non-moving parts of the animal for a perfectly spherical clear spot. At intervals it contracts and disappears, hence the name *contractile vacuole.* If you have a favorable specimen, watch and describe in your notes the behavior of the vacuole, and time the interval between contractions. When the vacuole reappears after a contraction, is it the same size as previously? Sometimes two or more vacuoles may be present, but they generally coalesce into one later. What is the function of the contractile vacuole (R)?

b) The nucleus: This is a sharply outlined, finely granular body in the neighborhood of the contractile vacuole, often in contact with it and of about the same size as the fully expanded vacuole. Its granules, which are chromatin granules, are much finer and more regularly distributed than those of the endoplasm. Determine the real shape of the nucleus by watching it as it rolls along in the moving endoplasm. Is it spherical? Occasionally specimens are found which have two nuclei.

c) Food vacuoles: The endoplasm usually contains particles or masses of digesting food, the larger of which may be inclosed in a drop of fluid and hence are called *food vacuoles.* Find out in your text how the *Amoeba* digests food. How does the process compare with that of the frog?

Make an enlarged drawing of the *Amoeba,* and put into it all of the details of structure described above. Indicate the granular appearance by stippling with the pencil point.

4. Activities.—Can you observe any indications that the amoeba is irritable and responds to the varying conditions of its environment? For instance, what does it do when it comes in contact with an obstacle or when other animals strike against it? If possible, observe the reaction to food material. How do you suppose the amoeba distinguishes between food particles and other particles? What does the amoeba do when it comes in contact with food material? Describe in your notes and sketch this behavior, if you are so fortunate as to observe it.

5. General considerations on the amoeba.—Does the amoeba carry on all the physiological processes that we found to occur in the frog? Does it have any special organs for these processes? What are the significant differences and resemblances between the amoeba and the frog? What does the frog gain by its greater complexity of structure? To what stage in development of the frog does the amoeba correspond? Give these questions careful thought and answer specifically in your notes.

C. PARAMECIUM

The study of this animal is usually somewhat difficult owing to its swift movements. In order, therefore, to obtain satisfactory results the student must follow the directions very closely (Hegner, chap. v, pp. 59–79).

1. General form and movements.—Mount a solid piece of scum and a drop of water from the culture on a slide and examine under the low power *without* a cover glass. A number of different organisms will probably be present and of these the relatively large, greenish, slipper-shaped animal██ ███. Note carefully the general shape of the body. Are the two en██ ███ form? Is one end always directed forward in swimming? If so, ███ Does the animal have a definite permanent shape?

Watch and describe in your notes the swimming movements. Obs███ ██ the animal when swimming in a free field revolves upon its long ████, ███ seems to swerve from side to side. What is the cause of the rotatio██ ███ swerving and what is the real path of the animal (see Hegner, p. 6█)?

By this time the student will probably have observed that the anim██ not symmetrical and cylindrical but that the anterior half is deeply ██████ as if a large slice had been cut out of it. This concave depression is call██ ██ *oral groove*. It is best seen by watching the animals as they revolve. The ██ of the *Paramecium* on which the oral groove is located is the ventral or ██ side; the opposite side, the dorsal or *aboral* surface; and right and left are ███ easily determined. (In Hegner, Fig. 23, p. 60, the labels *R* and *L* have ██ interchanged.) Observe the freely moving animals carefully until you have obtained a correct idea of the width, length, and direction of slant of the ██ groove. How wide is it compared with the anterior end? Does it slant fro██ right to left or left to right? (Remember that as the animal is transparent you cannot determine whether you are looking at the upper or under side of it, except by the use of the focusing screw.) How far posteriorly does it extend?

Make an outline drawing of the animal from the ventral side, about four inches long, showing its correct shape and proportions and correct appearance of the oral groove. Have the drawing approved by the assistant before proceed-ing. The details of structure are later to be entered upon this outline.

2. **Detailed structure.**—This study must be made with the high power, a proceeding usually fraught with difficulty for the student because the animal will not stand still. One or all of the following methods of quieting the animals must therefore be employed and will ordinarily prove successful: (1) Always mount the animals with a piece of scum or other solid material from the culture. They will generally remain quiet around this, or around air bubbles, or often near the edge of the cover glass. This is the most satisfactory method of study-ing *Paramecium*, as the animals remain in a normal condition throughout. Every student should try this first. (2) Withdraw the water from the slide by applying a piece of filter paper to the edge of the cover glass, so that not enough water remains for the animals to swim in. Under these conditions, however, the normal shape and appearance of the animals is lost, and they soon burst and die so that observations must be completed quickly. (3) Mount a drop or two of *Paramecium* on a slide and place in the center a *small* drop of a very dilute solution of formaldehyde. Cover. The animals near the formalde-hyde will soon slow down and finally become entirely motionless; those farther away will eventually also succumb. For a time, the motionless animals retain

the normal shape and structure (although the trichocysts are often discharged), but finally they round up and become abnormal. When this happens find another specimen. This method is very successful, *provided that the proper concentration of formaldehyde is used.*

All observations with the high power must be made with a cover glass over the material. Do not allow the material to dry up. Make observations only upon normal specimens, as far as practicable. With a little patience all of the following details of structure can be observed.

Paramecium, like *Amoeba*, consists of but a single cell, but whereas the latter is a mass of nearly homogeneous protoplasm, showing little differentiation, the protoplasm of *Paramecium* has become differentiated into a considerable number of different parts, as the following description will demonstrate.

a) Ectosarc: The outer layer of protoplasm of the *Paramecium* is called the ectoplasm or *ectosarc*. The surface of the ectosarc is differentiated as a firm membrane, the *cuticle* or *pellicle*, to which the animal owes its permanent shape. It is marked with a mosaic of hexagons which is not usually demonstrable (see Hegner, Fig. 24, p. 62). From all over the cuticle arise delicate threadlike processes of the protoplasm, the *cilia*, whose co-ordinated movement like numerous little oars propels the animal through the water. The cilia are practically of the same length over the body except that there is a tuft of longer ones on the posterior extremity. Those in the oral groove are especially active. The cilia are best seen by focusing on the edge of the animal. The layer of ectoplasm under the cuticle contains innumerable rodlike bodies lying parallel to each other and at right angles to the surface. They are best seen by focusing on the edge of the animal, as in the central parts of the animal they are viewed from the end and hence appear as dots. They are called *trichocysts* and appear to be little oval sacs containing a viscous fluid. Upon stimulation this fluid is discharged through the minute opening of the trichocyst and hence is squeezed out into the form of a long thread. **Enter** the layer of trichocysts and the cilia upon your outline drawing. Do not make a new drawing.

b) Endosarc: The central mass of the *Paramecium* is the endoplasm or *endosarc*. It is much more fluid than the ectosarc and is filled with granules. It generally contains a number of spherical vacuoles packed with food particles, designated as *food vacuoles*. It contains the nuclei (see below). **Enter** details of the endoplasm on your drawing.

c) Digestive apparatus: This is best studied on a normal active animal, one that is resting quietly near some object. It can also be seen on formalized animals but *not* on those which have been flattened out by withdrawal of water. Having found a favorable individual, watch the posterior end of the oral groove (be sure you know where this is) for an oval clear spot. This is the so-called *mouth*. Then observe a clear funnel-shaped curved cavity leading posteriorly from the mouth down into the endoplasm. This is the *gullet* or *cytopharynx*.

In favorable cases you may observe minute food particles shooting down the gullet and collecting into balls in its inner rounded end. Note the great activity of the cilia in the gullet, where they are fused into a vibrating membrane, known as the *undulating membrane*. The oral groove, mouth, gullet, and undulating membrane are the food-catching apparatus of the *Paramecium*. Near the gullet is a region inappropriately named the *anus*, where undigested food material is extruded. It can be seen only when material is passing out through it. (The terms "mouth," "pharynx," "anus," etc., should not have been applied to the *Paramecium* as they do not in the least correspond to the structures so named in the frog, except from the point of view of their function.) **Enter** the above-mentioned details on your drawing.

d) Contractile vacuoles: Observe a circular clear spot near each end of the animal. These are the contractile vacuoles. They are located between the ectosarc and the endosarc but firmly attached to the former. Are they on the ventral or dorsal side? Watch the contraction and note the radiating *canals* which appear like the petals of a flower around the point where the vacuole has disappeared. What is the average time between successive contractions of a vacuole? How does this compare with the contraction interval of the vacuole of *Amoeba*? The anterior vacuole of *Paramecium* generally contracts more frequently than the posterior one. (Note that the vacuoles may cease to contract or may contract very slowly in formalized specimens, and become very large and abnormal in behavior in flattened specimens.) **Enter** vacuoles and canals on your drawing.

e) Nuclei: Paramecium and its relatives have two nuclei, a large *meganucleus* (also called *macronucleus*) and a small *micronucleus*. They cannot be seen in normal animals, but the meganucleus can usually be recognized as a central large irregular mass in flattened and formalized individuals. Study the nuclei in the stained slides of *Paramecium* which are in your box. The meganucleus is a large-lobed and folded mass in or near the center, and the micronucleus is a small spherical body lying in a concave depression of the meganucleus. **Put the nuclei** in your drawing in their proper places.

3. **Experiments on Paramecium—**

a) Formation and course of food vacuoles: Start this experiment at the beginning of the laboratory period, as it requires some time for completion. It need not be watched continuously and the other experiments may be performed during the progress of this one. Mount some *Paramecia* with a piece of scum, add a drop of India ink suspension, and cover. This experiment can be carried out only on normal active animals, and the observations should be made on animals which are resting about the piece of scum. Do not let your preparation become dry. Find a quiet individual, and note that the particles of carbon are swept rapidly down the gullet to its rounded termination where they collect into whirling spherical masses. Observe that from time to time one of these masses breaks off and passes into the endoplasm as a food vacuole. Observe in what part of the body the food vacuoles first collect and where they are found later. Find out by observations at intervals their exact course in the endoplasm and **make a diagram** to indicate your observations. The movement of the vacuoles is of course due to a slow circulation of the endoplasm. You may, if you watch the preparation long enough, be able to observe the discharge of the carbon through the anus.

b) Discharge of the trichocysts: Mount some *Paramecia* on a slide, add a drop of picro-acetic acid, cover gently, and examine. Each animal will be found surrounded by a halo of long threads which are the discharged solidified contents of the trichocysts. **Make a drawing** showing relative lengths of cilia and the threads from the trichocysts.

c) The avoiding reaction: Mount some *Paramecia* as usual and observe with the low power. What happens when the animal strikes an obstacle? This reaction is called the *avoiding reaction*. **Make a diagram** to illustrate the reaction, showing the position of the *Paramecium* before, during, and after striking the object.

d) Reaction to chemicals: Obtain a considerable number of *Paramecia* and spread them out over the slide. Place the slide on the table and drop a small crystal of common salt in the center of the slide. Observe with the naked eye the behavior of the *Paramecia* toward the salt. What does each *Paramecium* do on coming into the neighborhood of the salt solution? What part does the avoiding reaction play in the behavior? **Make a diagram** to illustrate the results of this experiment.

4. Reproduction.—This process takes place by division (*fission*) of the animal into two halves by a transverse constriction. A sort of sexual act called *conjugation* also occurs at intervals, although it has been demonstrated that this is not·necessary for the continued existence of *Paramecium* (Hegner, p. 73).

a) Fission: In fission, both of the nuclei play an active rôle. The meganucleus divides by direct division; the micronucleus by a very primitive kind of mitosis, with the formation within it of fibers like the spindle fibers (see Hegner, Fig. 32, p. 70). Study and **draw** from the prepared slides **three stages** of fission, marked "early," "middle," and "late." Note that each slide contains only a few individuals which are dividing while the rest of the specimens are merely normal. The location of the dividing individuals has been indicated on most of the slides by ink marks. Do not draw until you are sure that you have found a specimen in fission.

(1) Early fission: The beginning of fission is recognizable by the elongation of the meganucleus, which becomes nearly as long as the animal. The micronucleus has moved away from its usual position in a depression of the meganucleus and is free in the cytoplasm, where it will be found in various stages of division. It accomplishes this division by pulling into two halves which move

farther and farther apart from each other, drawing out the connecting band into a thinner and thinner thread which finally ruptures. The division of the micronucleus is completed very early in the process of fission.

(2) Middle fission: The micronucleus is completely divided and one of its halves will be found near each end of the animal. The meganucleus is separating into two halves connected by a slender strand. A transverse constriction is present across the center of the cell.

(3) Late fission: The division of the meganucleus has been completed. The cytoplasmic constriction has divided the cell nearly or completely in two, but the two halves still cling together. Each half has regenerated those structures which it lacked. The whole process of fission requires in life about one half an hour. The daughter-cells at first smaller than the normal *Paramecium* soon attain adult size and proportions.

b) Conjugation: Look over the slide for cases in which two *Paramecia* are united by their oral grooves. **Draw.** It is not practical for beginning students to study the stages of the conjugation process but the texts should be consulted for this information (Hegner, pp. 68–73).

5. General considerations on the *Paramecium*.—Is *Paramecium* more differentiated than *Amoeba*? In your answer, compare in detail *Amoeba* and *Paramecium* as to definiteness of form, organs of locomotion, place of ingestion and egestion of food, etc. In what part of the protoplasm, ectoplasm, or endoplasm has the greater degree of differentiation occurred? Can you give a reason for this? What does *Paramecium* gain by its more complex structure?

D. GENERAL STUDY OF PROTOZOAN CULTURES

In order to obtain Protozoa in abundance it is necessary to cultivate them in a food-containing solution, which is called a *culture*. Such cultures are made by adding boiled grain, hay, bread, pond weeds, etc., to a considerable quantity of water; bacteria from the air fall into the culture, flourish there upon the food material which was put into the culture, and furnish food for the Protozoa, which are added to the culture from a natural source, such as pond water. Owing to the abundance of bacteria in such artificial cultures the Protozoa increase by division until enormous numbers of them will be present in a relatively small quantity of water.

Besides bacteria, yeasts, fungi, and other low plant forms and Protozoa, small multicellular animals also occur in the cultures, and a hasty study will be made of them also.

Obtain a drop or two of water from Protozoa cultures which are supplied; cover, examine, and identify as far as possible the forms present. Sketches of the more interesting ones are desirable if time permits. In the case of Protozoa particular attention should be paid to organs of locomotion and specializations of structure.

1. **Protozoa.**—Learn to distinguish Protozoa from other microscopic animals. They are recognizable by absence of cell walls and absence of organs, their bodies having a granular structureless appearance. The common Protozoa met with in cultures besides *Paramecium* are:

a) Vorticella: This relatively small protozoan is readily recognized by its bell-shaped body and the slender stalk arising from the top of the bell and permanently attaching the animal to other objects in the culture. The stalk is contractile; in fact it contains a spiral muscle, which on shortening draws the stalk into the shape of a spiral spring. We thus see that muscular fibrils can be differentiated even within the limits of a single cell. The free end of the bell bears a circle of swiftly vibrating cilia, and a large opening to the gullet. Look for food vacuoles, contractile vacuoles, etc. The horseshoe-shaped macronucleus can be seen only after staining with aceto-carmine.

b) Stentor: This animal is a large trumpet-shaped form, usually attached to objects. The broad end bears a circle of large cilia, and a conspicuous spiral gullet. The ectosarc is striped. The stripes are muscular fibrils which enable the *Stentor* to undergo considerable changes of shape. The macronucleus resembles a string of beads and is generally a conspicuous object.

c) Hypotrichous ciliates: This group of ciliates is easily distinguished by a jerky darting method of locomotion, and the possession of large cilia which are used as legs for creeping over objects. These cilia are really fused bundles of cilia, usually called *cirri*, and it is a remarkable fact that each one of them can be moved independently. Look for gullet, contractile vacuole, etc. The nucleus cannot be seen without staining. Use aceto-carmine. Common forms are *Euplotes, Stylonichia, Oxytricha.*

d) Other ciliates: Ciliates are the commonest Protozoa, and a great variety may be expected in cultures. *Frontonia* is similar to *Paramecium*, but larger and more oval in form; *Didinium* may be recognized from Hegner's Fig. 26 (p. 63); *Spirostomum* is a long, slender, cigar-shaped ciliate with oblique muscle stripes; *Lacrymaria* is spindle-shaped with a very long, slender, mobile "neck"; *Dileptus* is a large form with a short contractile neck; *Colpoda* is a small oval type with a marked indentation near the anterior end; *Coleps* is distinguished by an armor of small squarish plates; *Halteria* is a quite small, nearly spherical ciliate, moving by swift darts. These are the forms which we commonly get in cultures "seeded" from natural waters in the Chicago region, but many others may be expected.

e) Heliozoa, or sun animalcules: Spherical protozoans with stiff radiating pseudopodia and bubbly protoplasm are not infrequently found. They are related to Amoeba. *Actinophrys*, small, and *Actinosphaerium*, quite large, are the common kinds.

f) Amoeboid organisms: The only other common protozoan, similar to Amoeba, is *Arcella*. This animal resembles Amoeba in all essential respects except that it secretes a brown hemispherical shell in which it lives and which it carries on its back, so to speak, when it moves about. It is relatively common in cultures.

g) *Flagellates:* Organisms similar in appearance to the Euglena figured and described in Hegner (p. 82) are not uncommon in cultures. Some are green, some colorless, all possess long threadlike *flagella* as locomotor organs, although these are often very difficult to see, as they are transparent. Flagellates are best recognized by a peculiar swaying type of swimming movement, quite distinct from the movement of ciliates.

2. **Multicellular organisms.**—These are readily distinguishable from Protozoa by the presence of definite organs in their bodies. See Hegner (pp. 5–6) for the general characters of the different phyla. One usually finds in cultures the following types:

a) The *rotifers,* or wheel animalcules (Phylum *Trochelminthes*): These extremely common animals are at once known by the presence of one or two disks of cilia on their anterior ends. The constant movement of these cilia produces the illusion of a rotating wheel. The ciliary apparatus can be folded into the head. Other interesting features are the internal jaws, which keep up a constant chewing movement, the jointed telescopic posterior end, often provided with one or more "toes" for clinging to objects, and the hard case, or *lorica,* which often incloses the animal. The most common rotifer in cultures is *Philodina,* which possesses a pair of "wheels."

b) *Roundworms* or *nematodes* (Phylum *Nemathelminthes*), slender, cylindrical, wormlike animals, pointed at both ends, without cilia, moving by violent alternate curving and straightening of the body.

c) *Flatworms* (Phylum *Platyhelminthes*), slender flattened animals, moving in a smooth gliding manner, due to the presence of cilia; with more or less definite heads. *Stenostomum* is the commonest form, and often occurs in chains, produced by fission.

d) *Chaetonotus* (Phylum *Trochelminthes?*): This animal, possibly related to the rotifers, resembles a ciliate protozoan. It has a slender flexible body, possessing in front a rounded head region, followed by a short "neck," and at the posterior end two pointed processes. The whole surface is covered with short spines.

e) *Naids* (Phylum *Annelida*), slender cylindrical worms, provided with projecting bristles at regular intervals. The smallest ones belong to the genus *Aeolosoma,* and have beautiful red, orange, or green spots. Some (*Stylaria, Pristina*) have a long, slender, very active proboscis at the anterior end. *Dero* is provided at the posterior end with an expansible hood bearing several ciliated gills. Most of the naids are quite large compared to the other forms we have been describing. They also occur in chains produced by fission.

f) Water mites and *water bears* (Phylum *Arthropoda*): These are round to oval flat animals with projecting legs. The water mites have six jointed legs in the young state, eight in the adult. The water bear has eight short, non-jointed legs provided with claws.

g) Entomostraca (Phylum *Arthropoda*): These common animals have conspicuous jointed legs, and often large jointed antennae projecting from the head. They swim in a jerky manner. One of the commonest types is *Cyclops*, a small animal with a jointed body, single median eye, long swimming antennae, and slender terminal spines; often also with two sacs full of eggs hanging to the body. The *water fleas* have well-marked heads with large eyes, and powerful swimming antennae, but the body is inclosed in a double shell, from which the jointed legs may be protruded from time to time.

This brief account of the microscopic animals commonly found in pond water is merely intended to give the student an idea of the variety of form and structure of aquatic life. Students who are interested will find more detailed accounts of these animals in Stokes's *Aquatic Microscopy*, obtainable in the library.

X. PHYLUM COELENTERATA

A. HYDRA

1. **General structure.**—Note the appearance of the animals in the culture. They are *sessile*, that is, attached to the glass, plants, etc., by one end while the rest of the body hangs free in the water. Obtain a living *Hydra* and place in a watch glass with sufficient water to cover the animal. Examine with the lowest power of the microscope or with a hand lens, and note the following parts (Hegner, chap. viii, pp. 116–38).

The body or *column* of the animal is a cylindrical elastic tube, capable of great extension and contraction. From one end of this, which is the *oral*, *anterior*, or *distal* end, arise a number of radially arranged slender outgrowths, the *tentacles*. The number of these varies from four to ten, but is generally five or six. The oral extremity between the bases of the tentacles forms a conical elevation, the *hypostome*, in the center of which the *mouth* is located. The mouth when closed has a star-shaped appearance but is usually difficult to see, unless the hypostome happens to be turned directly upward. The flattened base of the animal is the *posterior*, *aboral*, or *proximal* end, is often designated as the *foot* or *basal disk*, and secretes a cement-like substance by which the organism attaches itself. Column and tentacles are hollow, inclosing a cavity known as the *gastrovascular cavity* because it has both digestive and circulatory functions. Under the low power the gastrovascular cavity appears outlined by brownish lines in most specimens; it is usually an extensive cavity only in the anterior half of the column, being reduced posteriorly to a slender canal.

Observe the tentacles more closely and note that they bear numerous protuberances, each of which contains a collection of very small oval sharply outlined bodies, the *stinging cells* or *nematocysts*. Such a collection of stinging cells is called a *battery*. At the end of each tentacle is a great mass of stinging cells. They also occur sparingly on the column.

Hydra is not bilaterally symmetrical, like the frog, where there is only one possible plane that will divide the animal into similar halves, but is *radially* symmetrical, that is the parts of the body radiate from a common axis so that a number of planes of symmetry could be passed through the animal. How do you think the differentiation and development of the anterior end compare with that of the frog? Does it seem to have any special structures different from the rest of the body?

Draw a *Hydra* in the extended state showing the above-mentioned details.

2. **General behavior** (Hegner, pp. 127–33).—On your own specimen or on those in the general culture jar perform the following simple experiments. Use a clean needle, touch the animals gently, and wait for the animal to expand completely before stimulating it again or use a different specimen. Touch one tentacle of a fully expanded individual gently. Does it contract? Do other parts contract? Touch two or three tentacles simultaneously. Is the response more marked than before? Try the comparative effect of a touch of the same intensity on the foot, middle of the column, and hypostome. What parts of the animal are the most sensitive? least sensitive? Try difference in response to a weak and strong stimulus applied to the same region. Stir the water in the culture and observe what happens. Do any of these experiments indicate that the stimulus is conducted from the point of stimulation to other parts, as in the frog? As in the frog the mechanism of irritability and perception of stimuli is a nervous system and sensory cells and the mechanism of response consists of muscle fibers. Owing, however, to the very simple structure of these systems, there is but one response to all kinds of stimuli, namely, contraction.

Make an outline drawing of a fully contracted *Hydra*.

3. **Cellular structure of the living animal.**—Mount a *Hydra* on a slide and support the cover glass with small bits of broken glass or slivers of wood so that the animal will not be crushed. The support must be thick enough to allow the animal to extend itself freely under the cover glass but not so thick as to interfere with the use of the high power. Small individuals are best for the purpose.

a) Cellular structure of the column: The animal must be fully extended for the following observations. *Hydra* consists of two layers of cells, an outer *ectoderm* and an inner *entoderm* with a structureless gelatinous sheet, the *mesogloea*, between them. It is thus like two closely fitting cylinders, one within the other. *Hydra* is therefore a *diploblastic* animal, that is, it consists of two *germ layers* (Hegner, p. 110). Examine the base of the column near the foot with the high power and try to see these two layers. First focus on the surface of the column and observe that it consists of a mosaic of elongated cells, with pointed ends. These are called the *epithelio-muscular* cells of the ectoderm because each has an epithelial and a muscular portion. A few nematocysts, oval distinct bodies, also occur lying in the ectoderm cells, and there may sometimes be observed between some of the epithelio-muscular cells greenish groups of very small cells, the *interstitial* cells. Now focus slowly downward, and a new layer of larger, more rounded, clear cells, the *nutritive muscular* cells of the entoderm, comes into view. Focus on the edge of the animal; the ectoderm then appears in profile as a rather thin layer, separated from the thicker entoderm by a dark line of uniform width, the mesogloea.

b) Cellular structure of the tentacle: Examine the base of a fully extended tentacle, and repeating the above-mentioned procedure observe first the ectoderm cells; then by focusing on the optical center note the large clear rectangular or cuboidal entoderm cells, the line of mesogloea, and the profile view of the ectoderm. The center of the tentacle is occupied by a slender canal, a branch of the gastrovascular system, but this is distinctly identifiable only when food particles in it can be seen moving back and forth. **Draw a small portion** to show the cell arrangement in optical section. Then examine the more distal (i.e., farther away from the body) portions of the tentacle to see the arrangement of the stinging cells. Each group of stinging cells, or battery, is contained within a single ectoderm cell, and causes a conical projection which appears clearly when the ectoderm is viewed in profile. Note the definite arrangement of large and small nematocysts in each battery and the projecting spines, the *cnidocils*, best seen in profile. **Draw a small portion** of the tentacle to show several batteries.

4. The nematocysts.—Remove the supports from under the cover glass and drum on the cover glass with the point of a pencil until the animal is thoroughly crushed. This discharges the nematocysts. With the high power examine undischarged and discharged nematocysts. There are usually three kinds of them, a large barbed and two small non-barbed varieties.

a) Large barbed nematocysts: In the undischarged state these are oval sacs, with one end flattened. From the flat end a hollow pouch projects into the interior, and from the inner end of this a coiled thread arises filling the rounded part of the sac. When discharged, pouch and thread are projected to the exterior, turning inside out. The pouch bears three large spikes and some quite small ones (see Hegner, Fig. 56, p. 121). Some kinds of *Hydras* have two sizes of this type of nematocyst.

b) Small oval nematocysts: These are much smaller than the preceding, oval or cylindrical, generally pointed at one end. When undischarged a spirally coiled thread fills the whole interior, projecting inward from the pointed end. When discharged the extremely long fine thread readily identifies this type.

c) Small spherical nematocysts: These are a little smaller than the preceding and more spherical. Each contains a thick thread which makes a single loop. inside the nematocyst and forms a tight little coil of three or four turns when discharged (see Hegner, Fig. 57, p. 122).

Nematocysts are not really cells but cell products secreted by the interstitial cells, which may frequently be seen clinging to them. Interstitial cells engaged in forming nematocysts are called *cnidoblasts*, and from them project slender spines, the *cnidocils*, whose stimulation is supposed to cause the explosion of the nematocysts.

Draw as many kinds of discharged and undischarged nematocysts as you can find without spending too much time upon them.

5. Cellular structure from slides.—

a) Longitudinal section: Examine slide " *Hydra*—long." Find a nearly median longitudinal section. Identify under low power the ectoderm, with its numerous darkly stained nematocysts, the line of mesogloea, and the elongated irregular highly vacuolated entoderm cells. In the region of the hypostome note the granular appearance of the entoderm, indicating glandular functions, as it here secretes mucus to aid in swallowing food; observe also the folds in this region, permitting great distension of the mouth. On some of the sections the continuation of the gastrovascular cavity into the tentacles can probably be observed. **Make a low-power diagrammatic drawing** of the section, constructing it from what you have seen on several sections.

b) Cross-section: Examine slide " *Hydra*—trans." Make a careful study with the high power of the cell structure.

(1) Ectoderm: The ectoderm consists of a sheet of cells one cell thick, forming a continuous outer layer of the animal, and appearing as the outer circle in the cross-section. It is composed mainly of large *epithelio-muscular* cells whose boundaries are often indistinct. The greater part of each of these cells is a polyhedral epithelial cell, but the base is drawn out into a long, slender, muscular fibril, running in a longitudinal direction. Thus the bases of all the ectoderm cells produce a longitudinal muscle coat for the *Hydra*, by means of which it is able to contract. (For the appearance of the muscle fibrils see below under "Mesogloea.") Each epithelio-muscular cell has a large nucleus, easily recognized by its black central nucleolus and distinct network. Within many of the epithelio-muscular cells are numerous darkly stained nematocysts, of which the various types described above will be recognized. Each is inclosed in its mother-cell, the *cnidoblast*, which is a modified *interstitial* cell. These latter are small, dark, granular cells, each about the size of a nucleus of the ordinary ectoderm cells, and occurring in groups or masses which are within or between the epithelio-muscular cells.

(2) Entoderm: The entoderm like the ectoderm is a continuous sheet of cells, forming a sort of inner tube for the animal. It is made up chiefly of the *nutritive muscular* cells, or *digestive* cells, large, elongated, vacuolated cells, with well-defined walls, and bulbous, crowded inner ends. They usually contain many food particles and droplets, have a nucleus like that of the ectoderm cells and a small amount of granular cytoplasm, confined mainly to the inner ends. Their outer ends, next to the mesogloea, are also prolonged into muscular fibrils, which run circularly, forming a circular coat, but this is relatively poorly developed as compared with the outer longitudinal coat. Situated between the inner ends of the digestive cells is another type of cell, the *secretory* or *glandular* cell, which produces the digestive enzymes. This appears as a small triangular cell, filled with a network, resting between the diverging inner ends of the adjacent nutritive cells.

(3) Mesogloea: The mesogloea is not a cell layer but a sheet of gelatinous material cementing the ectoderm and entoderm together. The longitudinally directed muscular bases of the ectoderm cells are imbedded in the mesogloea and appear there as dark dots.

Draw a small portion of the section in great detail to show all the kinds of cells present.

6. Reproduction.—*Hydra* reproduces by the asexual process of budding or by the reproduction of eggs and sperm in definite reproductive organs. Living material with sex organs is available only in the autumn.

a) Budding: If living specimens with buds are available obtain one and make a simple outline drawing.

b) Male reproductive organs: Examine slide at demonstration microscope. The male organs, *spermaries* or *testes*, form conical elevations in the ectoderm each provided with a nipple-like extension for the exit of the sperm. **Draw.**

c) Female reproductive organs: Demonstration slide. The *ovaries* are located nearer the aboral end than the testes, form low broad elevations, and lack the nipple. Each contains a single (or sometimes two) large amoeboid ovum. **Draw.**

7. General considerations on *Hydra*.—The chief differences that we may note between *Hydra* and the Protozoa is that it consists of several different kinds of cells, each with specific functions, and that these cells are arranged in definite layers, which foreshadow the systems and organs of the higher animals. To what stage of the embryonic development of the frog does the *Hydra* correspond? Which of its two layers is the more differentiated and what reason can you give for this? In *Hydra*, digestive, muscular, reproductive, and nervous (see Hegner, p. 125) systems are present, at least in a simple condition. What systems which the frog has are totally lacking in the *Hydra* and how does it accomplish the functions performed by those systems when they are present? In what way is the process of digestion in *Hydra* like that of the frog and in what way like that of the Protozoa (Hegner, p. 127)?

We may further notice that the *Hydra* is not merely a collection of many cells of several different kinds but that these cells act together for the general welfare of the animal, and each one is more or less helpless without the others, since it has become specialized for particular functions, and hence cannot successfully perform all functions like a protozoan cell. *Hydra* is therefore an *individual*, a unity produced by co-operation, ruled by a dominant region, the anterior end or head, which, to be sure, is as yet not very distinctly differentiated from the rest of the body.

B.   A COLONIAL COELENTERATE

For this study either *Obelia* or *Campanularia* may be used. As these animals live in the ocean, only preserved material can be obtained. Obtain a preserved colony and examine in a watch glass of water (Hegner, pp. 139–40).

1. General structure.—The coelenterate forms a branching colony of plant-like appearance consisting of a number of individuals, each of which is similar to a *Hydra*. Identify the rootlike basal branches, the *hydrorhiza*, by means of which the colony is attached to solid objects, the main stem, or *hydrocaulus*, and the branches, each one of which terminates in an individual which bears a general resemblance to Hydra, and is called a *zooid, hydranth,* or *polyp.* The whole colony is spoken of as a *hydroid colony.* The zooids are produced by budding from the stems, just as in *Hydra.* In fact if the buds of Hydra should remain attached to the parent a hydroid colony would be produced. **Make a small sketch** to show the general appearance of the colony.

2. **Detailed structure.**—Mount some of the preserved material on a slide or examine the slide of a hydroid in your box. Use the low power. The hydrocaulus and branches are covered by a horny layer, the *perisarc,* which they secrete. The perisarc extends up around each of the zooids in the form of a wineglass, the *hydrotheca.* At the base of each branch the perisarc is ringed, probably to make the stem more flexible. Within the perisarc is the *coenosarc,* a hollow tube comparable to the column of Hydra and consisting, like the latter, of ectoderm, endoderm, and mesogloea. Its cavity is the gastrovascular cavity and is continuous throughout the colony. There are two kinds of zooids.

a) Nutritive zooids: This is the more abundant kind of zooid. It consists of a cylindrical *body* from which arises the club-shaped *manubrium,* terminating in the wide mouth opening. At the base of the manubrium is a circlet of tentacles bearing nematocysts. The hydrotheca has a sort of shelflike extension inward at the point where the body of the zooid continues into the coenosarc. **Draw a nutritive zooid.**

b) Gonozooids: These modified zooids will be found where the branches arise from the hydrocaulus. Each is surrounded by a cylindrical case of perisarc, called the *gonotheca,* open at the end. The gonozooid consists of a central stalk or *blastostyle,* which is a degenerate nutritive zooid, and a number of saucer-shaped bodies borne upon the blastostyle and almost concealing it from view. The saucer-shaped bodies are *medusa buds.* They are formed by a process of budding from the blastostyle, and eventually they become free, escape from the gonotheca, and swim about in the water as little bell-shaped animals, called *medusae.* The medusa is the sexual individual and gives rise to eggs or sperms (on separate medusae); the fertilized egg develops into a new colony. Have the assistant show you a medusa. **Draw a gonozooid.**

In these colonial coelenterates there exist then three kinds of individuals: the nutritive zooid, whose function is that of food getting; the gonozooid, whose function is to bud off medusae; and the medusa, whose function is reproduction. Such a condition is known as *polymorphism,* or *division of labor,* because the organism instead of differentiating into organs for the performance of different functions, as the frog does (which its simple structure does not indeed permit), differentiates into several kinds of individuals for the performance of various functions. The hydroids also illustrate the principle of *metagenesis,* or *alternation of generations* (Hegner, p. 141), that is, the medusa or sexual generation is conceived of as alternating with the hydroid colony, or asexual generation. It would probably be advisable to drop this idea altogether and to regard the medusa simply as the final or adult stage in the development of the organism.

C. GENERAL SURVEY OF OTHER COELENTERATES

Study the collection of coelenterates provided for the class, reading what Hegner has to say about them (pp. 139–43). Acquaint yourself with the general appearance of the following chief groups of coelenterates:

1. **The hydroids.**—These animals are extremely common along the shores of the ocean, forming beautiful plantlike growths on rocks, wharves, plants, shells of animals, etc. Note various types of branching, size and shape of zooids, etc.

2. **Medusae.**—Medusae, as stated above, are the sexual stage of the hydroid colonies. They are small, gelatinous, bell-shaped animals with tentacles hanging free from the edge of the bell and a mouth dependent from the center of the concave surface of the bell. From the mouth, canals (usually four) radiate to the periphery of the bell and serve as a food-distributing system. A circular muscular shelf, the *velum*, extends inward from the edge of the bell; its contractions enable the animal to swim. The jelly-like composition of the medusae is due to the enormous development of the mesogloea.

3. **True jellyfishes.**—These differ from the medusae, which are also often called jellyfishes, through the absence of a velum, larger size and more saucer-like form, and more complicated structure. They usually do not pass through a hydroid stage.

4. **Siphonophora.**—These strange free-floating hydroid colonies illustrate the principle of division of labor carried to its highest development, since they may consist of seven or eight different types of individuals. The famous *Physalia*, or *Portuguese man-of-war* has a large "float" (which is a modified medusa), from the lower side of which the other members of the colony hang down, with long trailing tentacles. The *Velella*, or *purple sail*, has a flattened disklike float bearing an erect "sail," with zooids on the under surface of the float. In other types of Siphonophora the float is lacking but the upper part of the colony consists of a circle or a long chain of swimming bells, which are modified medusae.

5. **Sea anemones.**—These beautiful animals of the seashore are similar in appearance to hydroid polyps, but their internal structure is somewhat more complicated and they have no medusa stage. They have a stout column with an oral disk covered with tentacles.

6. **Stony corals.**—The common corals which build up the keys and reefs are small animals of the same structure as sea anemones. They secrete an external skeleton of calcium carbonate, which persists after they perish. In all dry specimens of coral the animals are of course destroyed, but the cuplike depressions

which they occupy in life are readily recognizable. When the cups are incomplete laterally, so as to be fused into long furrows, the familiar "brain" corals result.

7. **The horny corals** and their relatives include the coral used in jewelry, the organ-pipe coral, and other forms. The animals which secrete these skeletons have eight feathery tentacles.

8. **The sea feathers, sea fans, etc.**—These plantlike horny skeletons are built up by minute zooids, which occupy the small holes which will be found on closer inspection distributed abundantly over the skeletons.

A. PLANARIA

1. **General appearance and behavior.**—Obtain one of the living animals in a watch glass and examine with a hand lens (Hegner, pp. 153–58). Compare form, symmetry, and arrangement of parts with that of *Hydra* and the frog. Which of these two animals does it most resemble? What type of symmetry does it possess? Does it have anterior and posterior ends? dorsal and ventral surfaces? Is the head better differentiated than in *Hydra*? On the dorsal surface of the head is placed a pair of *eyes*, and each side of the head is extended into a blunt *contractile lobe*, or *auricle*.

Observe the peculiar gliding form of locomotion. To what is it due (*A*)? Does the animal exhibit muscular movements also? Do these suggest a better developed muscular system than is present in *Hydra*? How do the anterior end and auricles behave during locomotion? What does this suggest as to their function? Touch various parts of the animal with a needle and note its behavior and relative sensitivity of different regions. Do you think that *Planaria* is able to respond more effectively to stimuli and in a greater variety of ways than *Hydra*?

Make an accurate outline drawing several inches long of the animal.

2. **Detailed structure.**—Place a small *Planaria* on a slide and anaesthetize with a few grains of chloretone (*A*). After it has become motionless, turn it ventral surface up and cover with a cover glass, pressing gently so as to flatten the animal. If a cover glass is not heavy enough, use a piece of thin slide. The animal must not be crushed. Examine under the low power of the microscope.

a) General regions and surface of the body: The narrow clear border of the animal is the *ectoderm*, the central branching dark brown or gray material is the *entoderm*, and the lighter brown material between these two is the *mesoderm*. Focus on the surface of the body and note the numerous minute *pigment granules* which give the animal its dark color. Focus on the ectoderm with the high power, and observe the numerous rodlike bodies, placed at right angles to the surface, which it contains. These are called *rhabdites*, and when discharged on the surface they soften to produce mucus. The ectoderm is clothed with cilia, particularly on the ventral surface, and these may be seen by focusing on the edge of the ectoderm.

b) Sense organs of the head: Examine an eye. It consists of a clear area, where the sensory cells are located, and a crescent-shaped mass of very black, closely packed granules, whose function apparently is to concentrate the light upon the sensory cells by preventing reflection and diffusion. As the eye of

Planaria has no lens it cannot perceive objects but can only distinguish different degrees of intensity of light. The general body surface of *Planaria* is also sensitive to light but less so than the eyes. The pointed tip of the head and the auricles appear lighter in color than the rest of the body. They are important sensory areas, containing a variety of sensory cells, principally for contact and chemical sense, the latter essential to the detection of food.

c) *The digestive system:* In the center of the body is an oval clear region, the *pharyngeal chamber*, in which is located a cylindrical contractile tube, the *pharynx*. The pharynx is attached at its anterior end to the wall of the pharyngeal chamber, and its free posterior end has a wide opening. The pharyngeal chamber opens on the ventral surface by the *mouth*, which appears as a small circular clear area, devoid of granules, situated in the median line anterior to the posterior end of the pharynx. When the animal feeds, the free end of the elastic pharynx is protruded as a *proboscis* through the mouth. The student should understand clearly the relations of mouth, pharynx, and pharyngeal chamber and have the assistant make further explanations, if necessary.

At its anterior end the pharynx opens into the *intestine* or digestive tract proper (the pharynx being really an outgrowth of the body wall), which divides at once into three main branches, a median anterior branch extending forward as far as or between the eyes, and two posterior branches which pass backward, one on each side of the pharyngeal chamber, to the posterior end. Each main branch gives off numerous side branches or *diverticula*, often very irregular in form, and these may make secondary fusions to form further longitudinal branches. The digestive tract generally appears gray or brown and is easily seen if the animal is well pressed out and more than the usual amount of light admitted through the diaphragm.

Add to your outline drawing these details of structure.

The digestive canal of *Planaria*, although commonly called intestine, differs from the true intestines of the higher animals in that it consists of a single layer of cells, the entoderm, and is not separated from the other structures of the body as a definite tube. It is in reality much more comparable to the wall of the gastrovascular cavity of *Hydra* and its cavity corresponds completely to this cavity in *Hydra*, except that it is more branched and hence more effective as a food-distributing system.

d) *Other systems of Planaria:* The animal has well-differentiated nervous, excretory, and reproductive systems, but unfortunately these are difficult or impossible to study. Read carefully Hegner's description of them (pp. 155-57). In the nervous system note especially the concentration of nervous tissue in the head between the eyes, constituting the beginning of a brain. The structure of the excretory system should also be carefully studied, as it is constructed upon a primitive plan from which the excretory systems of nearly all animals, including the vertebrates, can be derived.

3. **Study of the microscopic structure.**—The slides contain sections in front of, through, and behind, the pharynx. All three sections should be made use of in the following examination.

The structure of *Planaria* differs from that of *Hydra* in that a third layer, the mesoderm, composed of muscles and connective tissue is present between the ectoderm and entoderm. *Planaria* is therefore a *triploblastic* animal, consisting of three germ layers. In the sections the ventral side is the flattened side; the dorsal side the convex side.

a) Ectoderm: This forms a thin layer of ciliated epithelial cells, the *epidermis*. In the epidermis may be distinguished rodlike bodies, the *rhabdites*, which when discharged swell and form mucus and *mucus* cells, containing granules which are also discharged to form mucus. These mucus cells are unicellular glands.

b) Mesoderm: Just under the epidermis is a thin layer of circular muscle cells. Within these is a rather indefinite layer of longitudinal muscle cells which appear as circles. There are also dorsoventral muscles, extending between the dorsal and ventral sides, between the branches of the digestive tract. The rest of the mesoderm forms a characteristic connective tissue, called *parenchyma*, which consists of branching cells and fills up all space between the ectoderm and entoderm so that a coelome is not present.

c) Entoderm: Each section contains several circular hollow masses which are the cross-sections of the digestive tract. The wall of the digestive tract consists of a single layer of elongated epithelial cells, which are the entoderm cells. These are generally so vacuolated and full of food material as to be almost unrecognizable.

d) The pharynx: The pharynx lies in a central cavity, the pharyngeal chamber. On some slides the ventral opening of this chamber, the mouth opening, may be present. The pharynx is an outgrowth of the body wall which has come to lie in a depression of the body wall, the pharyngeal chamber. The pharyngeal chamber is therefore lined with ectoderm, and the pharynx is covered and lined with ectoderm, the same as the outer layer of the body. The remainder of the pharynx is mesodermal, consisting of circular, longitudinal, and radiating muscle fibers, which the student should by this time be able to recognize, and parenchyma between these.

e) The ventral nerve cords: These can usually be recognized (best on sections anterior to the pharynx) as pale, round areas, near the ventral surface, about one-third to nearly one-half of the distance out from the median line. The pale portion consists of nerve fibers; often a few nerve cells, large cells with large nuclei, will be found along the dorsal border of the mass of fibers. **Make a diagram of the cross-section** through the pharynx.

4. **Feeding experiment.**—Demonstration. Observe the behavior of *Planaria* when meat is put into the pan. Do they recognize the presence of food? How? After the animals have attached themselves to the meat gently lift one

up and observe quickly the white protruded pharynx which is inserted into the meat.

5. **Regeneration.**—To save time and material members of each table may work together on the following experiment: Obtain several finger bowls and glass covers for them and wash both thoroughly. Take several *Planaria* and cut them up into pieces of various shapes and sizes. Cut some very small pieces. The animals are best cut by placing them on a glass plate, waiting until they extend to their full length, and then making cuts with a quick, firm stroke of a sharp scalpel. Place the pieces in the finger bowls with plenty of water, cover with glass plates, keep out of the sun, and change the water every few days. Examine at each laboratory period to observe the progress of the regeneration. **Make a series** of simple outline drawings to show the changes in form from day to day.

Is any piece of *Planaria* capable of regenerating a whole worm? Are any differences apparent due to differences in size or shape of the original pieces? Do very small pieces regenerate as well as larger ones?

6. **General considerations on *Planaria*.**—What systems does *Planaria* possess in common with *Hydra*? Are these better developed in *Planaria* than in *Hydra*? What system is present in *Planaria* which was entirely lacking in *Hydra*? What systems are absent in *Planaria* which the frog possesses? What type of tissue is present in *Planaria* which *Hydra* lacked? How do the muscle systems of the two animals differ? Is the digestive tract of *Planaria* a gastrovascular system as in *Hydra*, i.e., both digestive and distributive? Does *Planaria* still retain the protozoan method of digestion, as in *Hydra* (Hegner, p. 155)? What is the essential difference between cross-sections of *Hydra* and *Planaria*? Understand fully what are the corresponding layers in the two animals. To what stage in the development of the frog may the *Planaria* be compared?

B. GENERAL SURVEY OF OTHER FLATWORMS

Examine the specimens provided and familiarize yourself with the following groups of flatworms:

1. **Free-living flatworms or Turbellaria.**—These are slender, flat animals similar in appearance to *Planaria*, living in water or damp places, never of large size, often quite small. If living forms other than *Planaria* are available they will be demonstrated to you.

2. **Flukes or trematodes.**—These flatworms are external or internal parasites on other animals, particularly on vertebrates. They are broad, flattened, often leaflike animals, provided with suckers and hooks for adhesion to their hosts. The liver fluke, one of the largest forms, lives in the bile ducts of sheep and other domestic animals. The assistant should demonstrate living flukes parasitic in the frog. The commonest are the lung fluke, **Pneumoneces**, found inside

of the lungs of the frog; the bladder fluke, *Gorgoderina*, inside the urinary bladder; and *Clinostomum*, occurring in cysts in the peritoneum. Eggs and embryos of flukes frequently occur in the frog also.

3. Tapeworms or cestodes.—The tapeworms are degenerate flatworms living as internal parasites, commonly in the digestive tracts of vertebrates. They are extremely long, tapelike animals, as the name implies, with a small head, provided with hooks or suckers or both for clinging, and a body made up of a great number of joints, or proglottids. Each proglottid contains a complete set of the complicated reproductive organs, and when ripe they drop off from the posterior end of the worm so as to infect new animals.

XII. PHYLUM ANNELIDA

A. PRELIMINARY STUDY OF *NEREIS*

Nereis is an annelid living in the mud along the shores of the ocean. It is a relatively simple annelid and therefore used here to illustrate certain points. The dissection, however, will be done upon the earthworm, a more specialized form, but one more easily dissected.

1. **External anatomy of *Nereis*.**—Place a specimen in a dissecting pan. Compare general form and symmetry with animals previously studied. Is it bilaterally symmetrical? The most striking feature of the body is its division into a longitudinal series of rings, called *segments, metameres,* or *somites.* In this respect *Nereis* shows a marked advance over *Planaria* and resembles the frog. Each segment bears a pair of broad lateral outgrowths, the *parapodia,* by means of which the animal swims. Distinguish anterior and posterior ends, dorsal (rounded) and ventral (flattened) surfaces.

a) Head: The first segment of the body is markedly differentiated as a *head,* which bears a number of *sense organs.* The head segment consists of two parts, a dorsal squarish projection, the *prostomium,* which does not extend to the ventral side, and just behind this a complete, ringlike segment, the *peristomium,* which surrounds the mouth. (In most specimens the large pharynx will be found protruded from the mouth so as to extend some distance anterior to the head. This should not be confused with the true head.)

The sense organs of the head consist of *tentacles, palps,* and eyes. There are two short *terminal* tentacles projecting from the middle of the anterior edge of the prostomium. Lateral to these on each side is a thick jointed palp. On the dorsal surface of the prostomium are four eyes, indistinct blackish spots situated so as to form the four corners of a trapezoidal area. The two posterior eyes are likely to be concealed under the anterior edge of the peristomium. The eyes of *Nereis* are much more complex than those of *Planaria* and are provided with lenses, so that the animal can probably distinguish objects in a dim way. The peristomium bears four conspicuous, long, *lateral* tentacles, or *cirri,* on each side. It is probable that tentacles and palps are organs of touch, contact, chemical sense, etc. It is known that they are thickly strewn with sensory nerve cells.

The head of *Nereis* therefore represents another step in advance in the series of animal forms which we are considering in that it is more sharply differentiated from the rest of the body than is the case in the coelenterates and flatworms, and it is provided with a greater variety of sense organs and more complex and sensitive ones.

b) Posterior end: The last segment differs from the others in that it bears no parapodia. A central opening, the *anus*, is present, and two long tentacles, or *cirri*, project from the segment. The presence of the anus marks another advance over the previous forms in which the digestive tract has only one opening. In this respect again *Nereis* resembles the frog.

c) Remainder of the body: Note that with the exception of the first and last segments all the segments of the body are alike, not considering minor differences in the size and shape of the parapodia. This similarity of segments is not merely superficial but includes the entire internal anatomy. The same parts will be found repeated in each segment. *Nereis* is thus an ideal segmented animal.

d) Parapodium: Cut off one of these with a scissors, mount in water, cover, and examine with the low power of the microscope. The parapodium consists of several lobes, one of which, the *gill plate*, is very large and leaflike, and serves as a respiratory organ. In some relatives of *Nereis* it is transformed into a filamentous branched gill. Observe the stiff bristles, *setae* or *chaetae*, which project from the parapodium. They are characteristic annelid structures.

B. THE ANATOMY OF THE EARTHWORM

1. External anatomy (Hegner, chap. x, pp. 164–68).—Obtain a preserved specimen, place in the dissecting pan, and compare with *Nereis* as to form, symmetry, segmentation, etc. Are the two animals similar? Distinguish anterior and posterior ends, dorsal (dark-colored, rounded) and ventral (light-colored, flattened) surfaces. Is the head as well developed as in the case of *Nereis?* Is this related to the habit of life of the earthworm? The head of *Nereis* is to be regarded as a typical annelid head, while that of the earthworm is degenerate. Are parapodia present?

a) Head: The first segment is composed, as in *Nereis*, of a prostomium, the liplike process dorsal to the mouth opening, and the peristomium, the ring surrounding the mouth. The head of the earthworm probably consists of the first four or five segments, as indicated by experiments in regeneration. No eyes or other sense organs are present on the head (this being also associated with the habits of the earthworm) but numerous sensory cells are imbedded in the epidermis of the head and elsewhere.

b) Clitellum: Anterior to the middle of the worm a number of swollen segments occur, producing a distinct girdle around the body, called the *clitellum*. The clitellum is a glandular region which secretes the cocoon in which the eggs of the earthworm are laid. On the ventral surface of the clitellum is a pair of thickened ridges, the *tubercula pubertatis*. How many segments are there in the clitellum? How many segments anterior to the first clitellar segment? Is this number the same in all individuals? Remember that the prostomium is not a segment.

c) External openings: The mouth opening in the peristomium has already been noted. The last segment contains the anus. On the ventral surface of the fifteenth segment is a pair of conspicuous openings with swollen lips, the ends of the male ducts. The openings of the female ducts are on the fourteenth segment and much smaller than the male openings but not especially difficult to find on large specimens. There are also the openings of the two pairs of seminal receptacles on the ninth and tenth segments, a pair of excretory openings on the ventral surface of each segment, except the first three and the last, and dorsal openings, the dorsal pores, connected with the coelome, but all of these are minute and impossible to find.

d) Setae: The earthworm, like *Nereis*, bears setae which are arranged in four double longitudinal rows, two ventral and two lateral. Pass your finger up and down the earthworm and feel the setae as rough projections. With a hand lens make out their arrangement on each segment; the anterior part of the body is the most favorable place to see them. They are used by the earthworm to prevent slipping. **Make a diagram,** representing a segment as a ring and show the position of the setae on the ring.

2. **Internal anatomy** (Hegner, pp. 168–69).—

a) Structure of the body wall, coelome, mesenteries: The body is covered with an iridescent thin *cuticle* which is secreted by the epidermis. Strip off a small piece, spread out on a slide in a drop of water, cover, and examine with the high power. ·It shows striations at right angles to each other, which are the cause of the iridescence, and numerous pores, which are the openings of the gland cells of the epidermis to the exterior. **Draw** a small portion of the cuticle.

With a scissors cut through the body wall a little to the left of the median dorsal line from a point beginning about an inch behind the clitellum and extend the cut up to the prostomium. Be careful not to cut through anything but the body wall; if black material oozes out you may know that you are cutting into the intestine and should withdraw your scissors and begin again less deeply. Separate the edges of the cut and look inside. Observe that the *body wall* is separated from the *intestine*, the central dark tube, by a distinct space, the *coelome*. Observe further that this space is not continuous but is divided up into a series of compartments by delicate white partitions, the *septa* or *mesenteries*, which extend from the body wall to the intestine. What is the relation of these septa to the external rings? The coelome of the earthworm thus consists of a longitudinal series of chambers, which are paired, that is, one on each side, the intestine being inclosed between the two members of each pair. The inner walls of each pair of coelomic compartments therefore meet above and below the intestine to form *dorsal* and *ventral mesenteries*, although these are no longer complete in the adult worm. The anterior and posterior walls of each compartment come in contact with the walls of those in front of and behind it, producing double-walled partitions, the septa already mentioned. The outer wall of the compartment is fused with the inside of the body wall, forming a *parietal peritoneum*, while the inner wall is similarly attached to the intestine, forming a *visceral peritoneum*. The relations therefore of the coelome and its linings are identical with those that we found in the frog, except that the segmental septa are absent in the latter.

Cut through the septa so that you can lay out the body wall. In this operation it is best to stick a pin through the body wall on the left side; then lift up the other side of the cut and cut through the septa and pin out the right side; then move forward a short distance and repeat. Do not pull the wall apart but cut the septa with a fine scissors. Stick the pins in obliquely and firmly so that they will not get in your way and will not come out. Open up the body to the prostomium. Note the iridescent membrane, the *peritoneum*, which lines the body wall. *Cover the animal completely with water.*

Before beginning a detailed dissection of the various systems, the following conspicuous structures should be identified. The center of the body is occupied by the long brownish intestine. In the tenth, eleventh, and twelfth segments are pairs of large white bodies. These are the *seminal vesicles*, part of the male reproductive system. In the ninth and tenth segments are pairs of small white spherical bodies, the *seminal receptacles*, part of the female reproductive system. Loose white objects on each side of the intestine along its entire length are the excretory organs, or *nephridia*.

b) The circulatory system: This system is difficult and unsatisfactory to make out in dissection owing to the small size of the vessels, and the student is referred to Hegner (pp. 172-75) for details. The following parts should be noted:

(1) The *dorsal blood vessel:* This is the dark brownish line running along the median dorsal surface of the intestine. In many specimens the two pairs of branches which it receives in each segment from the wall of the intestine can be seen. Its branches from the body wall (torn off, of course) may also be found sometimes. Look for these.

(2) The *hearts:* Remove the seminal vesicles from the left side. Note that in the region where they were located and for some distance anterior to this the septa are stronger and more conspicuous than elsewhere, projecting out from the intestine like wings. Look in the eleventh segment, between the places occupied by the second and third seminal vesicles, for a pair of stout tubes arising from the dorsal vessel and extending ventrally. They are often dark-colored because of contained blood. Then look in each segment forward from this up to the sixth for a similar pair of tubes, grasping the winglike septa with a forceps and pulling them backward. These branches of the dorsal vessel, found in the seventh to the eleventh segments, are contractile tubes, inappropriately called *hearts*. The first pair is quite small. More than five pairs are sometimes found. Follow the dorsal vessel forward as far as you can beyond the region of the hearts.

(3) The *ventral blood vessel:* Loosen up the intestine completely on the left side so that it can be pressed over to the right side to enable you to see the median ventral line. Look here on the underside of the intestine for a brownish line, the ventral blood vessel. It is easiest to see under the dilated portion of the intestine behind the region of the seminal vesicles. Look for branches of the ventral blood vessel to the intestine, body wall, and the nephridia. Trace the connection of the hearts with the ventral blood vessel and follow the latter forward as far as you can.

(4) The *subneural vessel:* Directly under the ventral blood vessel is a white line, the *ventral nerve cord.* Cut this through at some convenient point back of the region of the hearts, pull up a short strip of it backward, and look on the underside for a longitudinal vessel looking like a dark line running down the middle of the white cord. Its branches from the nephridia and the body wall may sometimes be found. It is called the *subneural* vessel.

(5) The *intestino-tegumentary vessels:* Extending from the tenth segment forward, an extra vessel will be found on each side of the wall of the intestine. Hearts and other structures should be removed to see it.

Make a drawing from the side of the parts of the circulatory system which you have been able to find. Some specimens will show more of the vessels than others.

On living earthworms observe the rhythmic contractions of the dorsal vessel. In which direction do the contractions travel? Read Hegner (p. 175) on the course of the circulation in the earthworm.

c) The digestive system: It consists of the following parts (Hegner, pp. 169–71):

(1) *Buccal pouch:* This is a small tube extending from the mouth through about three segments. White, slender, muscle fibers, the *dilators* of pharynx, attach it to the body wall and allow of its expansion. On the dorsal surface of the buccal pouch, just before the pharynx begins, note the small, white, bilobed *brain.*

(2) *Pharynx:* This is the thick-walled portion following the buccal pouch. The muscular fibers attaching it to the body wall are much more numerous than in the preceding division.

(3) *Esophagus:* This long, slender tube extends through the region occupied by the hearts and seminal vesicles. These should be removed from the left side, if not already done, so that it can be clearly seen. In its wall in the region of the seminal vesicles are three pairs of brownish projections, the *calciferous glands,* said to secrete calcium carbonate into the digestive tract.

(4) *Crop:* The esophagus leads into the large thin-walled *crop,* which acts as a storage reservoir for food.

(5) *Gizzard:* Immediately posterior to the crop is the very thick-walled *gizzard,* in which the food is ground to small particles.

(6) *Intestine:* From the gizzard the brown thin-walled intestine extends straight posteriorly to the anus. Its brown color is due to peculiar cells, the *chlorogogue* cells, which cover it. Cut open the intestine for a short distance along one side, wash it out and observe a thickening in its median dorsal wall. This thickening, the *typhlosole,* is a longitudinal infolding of the dorsal wall, apparently for the purpose of increasing the absorbing surface of the intestine, for which function it is still further folded by transverse grooves.

Draw, showing all parts of the digestive system in relation to the segments, properly numbered. This drawing may be combined with that of the circulatory system, if desired.

d) The excretory system (Hegner, pp. 175–76): The excretory system of the earthworm consists of a pair of coiled tubes in each segment (except the first three and the last). Each tube is called a *nephridium.* The main part of each nephridium consists of a tube coiled transversely in the coelome of the segment and opening to the exterior of that segment by a *nephridiopore.* The beginning of the coiled tube is situated, however, in the segment anterior to the one where the main structure is located; the tube penetrates the septum and opens into the coelome of the preceding segment by a funnel, called the *nephrostome.* The nephridia are readily recognizable as the white loose structures on the sides of the digestive tract. Extend your original incision about one inch backward, *this time exactly in the median dorsal line.* Open out the wall as before by cutting the septa. Pull out the intestine. Pin out the body wall tightly. Observe with the hand lens the paired nephridia in each segment. By manipulating the septa you should have no difficulty in seeing the tube penetrating the septum and ending in the segment in front by the funnel, appearing as a white spot. Cut out a nephridium as completely as possible, mount on a slide in water, and cover. Examine with the low power. Identify the nephrostome (see Hegner, Fig. 89, p. 176); the thin tube which passes through the septum, and forms the first slender loop of the coiled portion; and the remaining wider loops.

e) The reproductive system: The dissection of the reproductive system is rather difficult and should be attempted only on large and well-preserved specimens. By constant reference to Hegner (Fig. 93, p. 181) and by exercising the utmost care in dissection the student will probably be able to find the parts of the reproductive system. As in *Planaria,* a complete set of both male and female organs occurs in the earthworm, that is, it is *hermaphroditic.*

The right side of your specimen has been preserved intact for this dissection. Remove the intestine carefully from the region of the seminal vesicles (no farther) and study their arrangement. The first pair are rounded sacs projecting forward into the ninth segment above the first pair of seminal receptacles. The second vesicles are more elongated and folded and occupy the eleventh segment. The third vesicles are the largest and are much folded, occupying the twelfth and part of the thirteenth segments. The three pairs are united by median masses located in the tenth and eleventh segments. Locate the fifteenth segment and stick a pin into it as a landmšrk. Gently press the vesicles and intestine to the left and pull off the nephridia from the right ventral body wall by grasping their free ends. Stretch out the right ventral wall tightly. Locate on it the ventral edge of the dorsal longitudinal muscle band, extending as a longitudinal line. Within and in contact with this will be found the lateral setae, projecting inward as minute white elevations, with a slender muscle band between the two members of the pair. One-third of the way between the row of setae and the ventral nerve cord find a white line extending from the middle of the fifteenth segment forward to the tenth. This is the *vas deferens* or male duct. Examine the septum between the fourteenth and thirteenth segments. The white object in it is the *oviduct* with a slender tube extending into the fourteenth segment. Then examine the septum between the twelfth and thirteenth segments. A minute white body, the *ovary*, will be found attached to it. The attached end is the broadest; the free slender pointed end projects backward into the thirteenth segment. If you are in any doubt that you have the ovary, remove it and examine it with the microscope, and the presence of round eggs, the largest of which are in the pointed extremity, will settle the matter. Next look in the twelfth segment for a slender branch extending from the vas deferens over to the fused median portion of the seminal vesicles. Dissect off the roof of the median portion of the seminal vesicle in the eleventh segment and note within its cavity the greatly folded *seminal funnel*. Anterior to this, attached to the septum between the tenth and eleventh segments, will be found by cautiously picking away the bases of the seminal vesicles a small but distinct round body, the *testis*. Repeat the directions in the three preceding sentences in the eleventh and tenth segments, and find the anterior branch of the vas deferens, the anterior seminal funnel, and the anterior pair of testes. The mother-cells of the sperm are produced in the testes and set free at an early stage into the seminal vesicles, in which they develop into sperm. Take out a small piece of the seminal funnel, put on a slide in a drop of water, cover, and mash by pressing on the cover glass with the finger and rotating it. Examine with the high power and note the myriads of slender-tailed *spermatozoa*. The two rounded *seminal receptacles* in the ninth and tenth segments were previously noted. They receive the sperm from another animal during copulation and are therefore parts of the female system.

Draw the reproductive system.

f) The nervous system: Remove the digestive system as far forward as the pharynx (Hegner, pp. 176–80). The ventral nerve cord, a white cord in the median ventral line, has already been identified. Clean away all tissue conceal-ing it. Examine it with a hand lens and note the enlargements or *ganglia* which are present in the cord in the middle of each segment and the *lateral branches* which arise from it. Cord and ganglia are really double, formed by the median fusion of two originally separate cords (as in *Planaria*), but the double character appears externally only in the anterior part of the system. Trace the nerve cord anteriorly. Loosen the pharynx and turn it forward. Find underneath it the anterior termination of the ventral cord in a large double mass, the *sub-esophageal* ganglia. Locate the brain on the dorsal side of the buccal pouch and cut away the pharynx and as much of the buccal pouch as you safely can, leaving the brain in place. Note the two cords which extend dorsally from the sub-esophageal ganglia to the brain, forming a circle through which the buccal pouch passes. These cords are called the *circumesophageal connectives*. Examine the brain and note that it consists of two distinct lobes, the *supra-esophageal* ganglia. Find the nerves running forward from the brain to the prostomium and from the connectives to the ventral portions of the first segments.

Draw the nervous system.

Strip off a piece of the ventral nerve cord, preferably near the anterior end, mount it in water on a slide and examine with the low power. Find out how many lateral nerves pass out from each ganglion and how many arise between the ganglia, and put this in your drawing. You will probably notice the subneural blood vessel and its branches on the ventral surface of the cord. Do not confuse these with nerves; they are generally yellowish and hollow, while the nerves are white and longitudinally striped.

g) Cross-section of the earthworm: Make a straight, clean cut across the earthworm in the center of a segment, that is, just halfway between two rings. Use the posterior part of the worm which has not been opened. Make another cross-cut near the first one so as to separate off a small section of the worm. Wash out the contents of the intestine in the piece and place it under water with the first cut surface up. Examine with a hand lens. Identify in the body wall the outer white *epidermis*, and under this the thicker greenish layer composed of *longitudinal muscles*. Observe in this the four pairs of setae projecting inward dividing the longitudinal muscle coat into *bands*. Of these identify the *dorsal* bands, extending from the median dorsal line to the lateral setae, the *lateral* bands between the lateral and ventral bundles of setae, and the *ventral* band across the ventral side. Slender bands also exist between the two members of each pair of setae. In the intestine identify the typhlosole. Between the intestine and the body wall in the coelome note the long white nephridia, opening to the exterior just outside the ventral setae. In the median ventral line above the ventral muscle band will be found the white section of the nerve cord. Gently lift out the nephridia and observe the septum stretching across the coelome.

Make a drawing of the cross-section.

3. Microscopical structure of the earthworm.—Examine slide "Lumbricus" (Hegner, Fig. 85, p. 166).

a) General appearance of the cross-section: Examine with the low power and identify the thick body wall, the coelome between the body wall and the intestine, and the intestine with its dorsal infolded typhlosole and outer covering of peculiar large cells. In the body wall, the most conspicuous layer is the feathery layer of longitudinal muscles, which is interrupted in eight places, four lateral and four ventral for the insertion of the setae. These interruptions divide the longitudinal muscle coat into bands, which were noted in the preceding paragraph in the cut surface of the earthworm and should be identified again. Identify the dorsal blood vessel lying above the typhlosole imbedded in the large pear-shaped cells; the ventral blood vessel below the intestine to which it is attached by a ventral mesentery; and the ventral nerve cord just below the ventral blood vessel. In the coelome may be seen traces of nephridia, blood vessels, and septa. **Make a diagram of the section.**

b) Structure of the body wall: Examine with the high power, and study the following layers:

(1) *Cuticle:* The outermost layer, a thin non-cellular covering.

(2) *Epidermis:* This contains about four kinds of cells. The most conspicuous are the *gland* cells, large elliptical cells filled with granules, which are forerunners of mucus. These gland cells open to the surface by way of the pores already noted in the cuticle. Between the gland cells are *interstitial* cells, elongated cells with broadened ends. At the bases of the gland and interstitial cells may be noted some small cells, which are sometimes considered to be a second layer. The fourth kind of cell, sensory cells, cannot be seen without special methods of staining (Hegner, Fig. 86, p. 167).

(3) *Circular* muscle layer: This consists of very long slender muscle cells imbedded in connective tissue, running circularly.

(4) *Longitudinal* muscle layer: This consists of muscle cells like those of the circular layer, but as they run longitudinally they appear in cross-sections as circles or ellipses. These longitudinal muscle cells are mounted on plates or septa of connective tissue, which extend out from the body wall at right angles to it. This arrangement of the muscle cells along these septa gives a feathery appearance in cross-section. The interruptions of the longitudinal muscles for the setae have already been noted.

(5) *Peritoneum:* The body wall is lined with peritoneum, a very thin layer of cells applied to the inner surface of the longitudinal muscle layer.

(6) *Setae:* If setae are present on your slide note their appearance. They are slender, yellow, curved rods of the same composition as the cuticle and like it secreted by the epidermis. The epidermis turns in to form a sheath around each seta and a sort of cap at the inner end of the seta, where the seta is secreted. From the epidermal sheath and cap muscles may be seen extending to the circular muscle layer. These muscles serve to move the setae in all directions.

Draw a small portion of the body wall in great detail.

c) *Structure of the intestinal wall:* Examine with the high power.

(1) *Chlorogogue* cells: The outer layer of the intestine consists of very large, irregular, somewhat pear-shaped cells which are modified peritoneal cells. They are named the *chlorogogue* cells.

(2) *Longitudinal* muscle layer: At the inner ends of the chlorogogue cells a row of circles, cross-sections of a very thin layer of longitudinally arranged muscle cells, appears.

(3) *Circular* muscle layer: A thin layer of circularly arranged muscle cells lies just within the preceding.

(4) *Vascular* layer: The blood vessels of the intestine, branches of the dorsal and ventral blood vessels, run in a definite layer in the wall of the intestine just within the circular muscles. Here they form a rich network, appearing in the section as irregular spaces, uniformly stained.

(5) Lining *epithelium:* The innermost layer of the intestinal wall, lining the cavity, is an epithelium, consisting of long, slender, ciliated, epithelial cells. This layer is almost as wide as the layer of chlorogogue cells. The nuclei are near the bases of the cells.

Draw.a portion of the intestinal wall in great detail to show these layers.

d) *Structure of the nerve cord:* Examine with the high power. The nerve cord is somewhat oval in outline and is covered externally by a sheath consisting of peritoneum, connective tissue, muscles, and blood vessels. Of these the more conspicuous ones are the *subneural* vessel on the median ventral side and the paired *lateral neural* vessels, one on each side of the subneural vessel.

The cord is more or less distinctly divided into two halves by a median partition. On the dorsal side of the cord are three large clear areas, the *giant fibers,* believed to be nerve fibers which run for long distances in the worm. Each is surrounded by a thick sheath. In the lateral and ventral portions of the cord may usually be seen several large pear-shaped nerve cells, each with a large nucleus and nucleolus. The nerve cells are more abundant if the section has happened to pass through one of the ganglionic swellings of the cord. The rest of the cord consists of cross-sections of nerve fibers, appearing like an open network.

Draw the section of the nerve cord.

e) *Longitudinal section of the earthworm:* Study and identify the parts already seen in the cross-section. Note particularly the septa.

4. General considerations on the earthworm.—What system is present in the earthworm which was lacking in *Planaria?* What systems which the frog possesses are lacking in the earthworm? How do you think the earthworm compares with the frog with respect to differentiation of organs and systems? What particular system shows the least advance over the same one in *Planaria* and least resembles that of the frog? In what ways does the digestive tract of the earthworm differ from that of *Planaria* and resemble that of the frog? In the cross-section of the earthworm what striking differences from the cross-section of *Planaria* are apparent? Are the layers of the body wall of the earthworm the same in general as those of the body wall of the frog? those of the intestine of the two animals? In comparing the cross-sections of *Hydra*, *Planaria*, earthworm, and frog one may note the following points: the gastrovascular cavity of the two former animals corresponds to the cavity of the intestine of the two latter; the lining epithelium of the intestine in earthworm and frog is the entoderm; the epidermis, the ectoderm; and all of the tissue in between is mesoderm.

C. GENERAL SURVEY OF ANNELIDS

1. **Polychaetes.**—The polychaetes are the group of marine annelids to which *Nereis* belongs. They are characterized by the possession of parapodia. Examine the various specimens available. Note form and segmentation of the body, development of parapodia, presence and location of gills (slender respiratory processes, often arranged in clumps), degree of development of the head. Many of the polychaetes live in tubes secreted by themselves, and hence the parapodia and head regions are often degenerate.

2. **Leeches.**—Leeches are common annelids of fresh water, readily distinguished by the presence of two suckers, one at each end of the body. Examine the specimens. Are parapodia or setae present?

XIII. PHYLUM ARTHROPODA

A. THE ANATOMY OF THE LOBSTER (OR CRAYFISH)

For this study either the lobster or the crayfish may be employed, although the former is preferable because of its greater size. The slight differences between the anatomy of these two animals are noted in the course of the outline (Hegner, chap. xi, pp. 193-225).

1. **External anatomy.**—Obtain a preserved specimen and place in a dissecting pan. The animal has a hard external covering, the *exoskeleton*, which corresponds to the cuticle of the earthworm, and it is a secretion of the ectoderm. It is composed of *chitin*, rendered hard by the deposit of calcium carbonate in it. Identify anterior and posterior ends, dorsal and ventral surfaces. Is the animal bilaterally symmetrical? Is it segmented? Is it clearly segmented along the whole axis of the body, like the annelids? What part of the body is most evidently segmented? what least?

The body differs greatly from the other invertebrates studied and resembles the frog in that it is divided into definite regions, the *head*, *thorax*, and *abdomen*. Head and thorax are, however, more or less fused into one region, called the *cephalothorax*. The single piece of the exoskeleton which covers the cephalothorax dorsally and laterally is called the *carapace*. A groove runs across the mid-dorsal region of the carapace and obliquely forward on either side. This is the *cervical groove* and separates the head in front from the thorax behind. Segmentation has been lost on the dorsal side of the cephalothorax through fusion of segments.

Another striking difference between the lobster and the other invertebrates previously studied is the presence of jointed appendages. Each segment of the body is represented by a pair of appendages, and it is thus possible to determine the number of segments even where the lines between them have been lost by fusion. In many arthropods, however, some of the appendages have been lost also, so that the determination of the number of segments in the body is sometimes a matter of great difficulty.

a) The head: The head terminates anteriorly in a spiny pointed projection of the carapace, the *rostrum*. The head is provided with a number of sense organs, exceeding, in variety and complexity, those of the lower invertebrates. There is a pair of large, stalked, movable *eyes*, which are probably not appendages. In front of the eyes occurs the first pair of appendages, the *antennules*, short, forked filamentous outgrowths. Just below these is the second pair of appendages, the *antennae*, long, flexible, many-jointed structures. Both antennae and antennules have tactile and chemical functions. On the ventral side of the head are additional appendages belonging to the head, surrounding the mouth and used for tasting, handling, and tearing food. These will be studied later.

b) *The thorax:* Two longitudinal lines on the dorsal surface of the thorax divide the carapace into a median *cardiac* region, covering the heart, and two broad, curved, lateral *branchial* regions, which cover the chamber containing the gills. The ventral side of the thorax bears five pairs of walking legs, of which the anterior pair is modified into large pincers; some of the appendages in front of the pincers also belong to the thorax.

c) *The abdomen:* The abdomen is plainly divided into seven joints or segments, six of which bear appendages, used for swimming. The first pair of abdominal appendages in male and female lobsters and in female crayfishes and the first two pairs in male crayfishes are modified for sexual purposes. The last pair of abdominal appendages is greatly broadened and forms with the last abdominal segment a broad swimming fan. The last segment, or *telson*, has no appendages and bears the anus on its median ventral surface. There is some doubt that the telson is a true segment.

d) *Study of a typical segment and pair of appendages:* As the abdominal segments are more distinct and less modified than the other segments, one of them, as the third or fourth, may be selected for study. Such a segment has the general shape of a ring, as in the annelids. Its exoskeletal covering may be divided into a convex dorsal portion, the *tergum*, a thin lateral plate, the *pleuron*, extending free ventrally into a point, and a slender ventral bar, the *sternum*. The region between the pleuron and the base of the appendage receives the name of *epimeron*. Between successive segments occur thin *arthropodial membranes*, where the calcareous deposit is lacking; these are best seen on the ventral surface between successive sterna, and they permit movement of the segments upon one another. Examine the joints between segments at the pleura; determine how they work by bending and straightening the abdomen.

Between the base of the pleuron and the sternum of the segment is a round area, into which the appendage is fastened by means of an arthropodial membrane. Cut through this membrane and remove a complete appendage. It has the following parts: It springs from the segment by a basal stem, the *protopod*. This really consists of two joints, a very small basal piece, the *coxopod*, and a long distal part, the *basipod*. From the basipod arise two flattened, many-jointed plates. The outer of these is the *exopod;* the inner one, next the median line, the *endopod*. **Make a diagram** of an ideal cross-section of the segment with its pair of appendages.

This type of appendage found on the abdominal segments is called the two-forked or *biramous* appendage. It is supposed to be the primitive arthropod appendage, all other kinds of appendages found in arthropods being derived from it by modification. Theoretically, the original arthropod consisted of a series of segments, like the abdominal segments of the lobster, all bearing biramous appendages.

e) Comparative study of the appendages: The appendages will be studied by comparing them with the typical biramous appendage described in the preceding section, and determining what modifications have occurred.   Consult Hegner, Table VII (pp. 197–99).

(1) The abdominal appendages: These are designated as *swimmerets*, since they are employed in swimming.   In the female they also serve as places of attachment of the eggs (Hegner, p. 213).   The swimmerets of the second to fifth abdominal segments are very similar to the typical appendage already described.   The sixth pair of swimmerets, called the *uropods*, is greatly enlarged and forms with the telson a powerful swimming organ, the *tail fan*.   Determine other differences between the uropods and the other swimmerets.   The first pair of swimmerets in the female lobster and crayfish is greatly reduced but otherwise similar to the others.   The first pair of swimmerets in the male lobster and the first and second pairs in the male crayfish are modified for the transference of the sperm to the female.   They consist of protopod and endopod fused into a hard, pointed, grooved structure.   Exopod is lacking, except in the second pair of the crayfish, where it appears as a soft, slender, lateral process.

(2) The thoracic appendages: Remove the branchial region of the carapace from the *left* side by lifting it up and cutting away the free portion.   The *branchial chamber* is thus exposed.   Observe that all of the thoracic appendages have gill-bearing processes extending up into the branchial chamber.   The thoracic appendages comprise five pairs of walking legs, called *pereiopods*, and three pairs of smaller appendages, called *maxillipeds*, anterior to them.

Remove completely the left *third* maxilliped, the appendage just in front of the large pincers.   Be sure to get the gill-bearing process with it.   Cut through the arthropodial membrane at the base and gently detach the appendage.   The basal joint, the coxopod, of the third maxilliped bears a delicate, hairy, leaflike expansion, the *epipod*, to which a feathery *gill* is attached.   Notice the great freedom of movement of the coxopod and explain.   The next joint distal to the coxopod is the basipod.   From it arise two branches, an inner endopod, consisting of five joints, and an outer exopod, of many small joints.   The maxilliped is therefore a biramous appendage, similar to the swimmeret, but showing a process of reduction of the exopod.   The basal joint of the endopod has a row of hard teeth, used in crushing food.

Remove the left second pereiopod with all of its parts and compare with the third maxilliped, placing both before you in the same position.   The pereiopod has a coxopod with an epipod and gill, a basipod, and an endopod of five joints. The comparison shows, however, that the exopod is completely lacking, and the pereiopod is therefore a *uniramous* appendage.   Observe the pincer at the end of the pereiopod and determine how it arose.

Examine the other pereiopods, comparing them with the second one and noting differences. Do all have pincers, epipods, gills? By moving the legs, note how the gills are moved. In what directions can the legs be moved upon the body? In what directions can the joints of the leg be moved upon each other?

The first pair of pereiopods is greatly enlarged, with a powerful pincer, or *chela*. Note that the two chelae of the lobster are not alike, but one is massive with broad crushing surfaces, called the *cracker claw*, and the other sharper and more slender, with little teeth, known as the *toothed claw*.

On the inner side of the coxopod of the third pereiopods of the female find the *female genital openings*. The *male genital* openings are in the same place on the fifth pereiopods.

Draw the third maxilliped and the second pereiopod in the same positions.

Next remove the left *second* maxilliped complete. Compare with the third maxilliped. Does it have the same parts? Notice reduction of the epipod and gill and the presence of only four joints in the endopod.

Remove the *first* maxilliped, and compare with the other two. Does it have an epipod? gill? Observe particularly that the appendage has become more broadened and leaflike. This is due to the moving of the coxopod and basipod from their original basal position to form the medial side of the appendage. The endopod is thus shoved laterally. Endopod and exopod of the first maxilliped are slender processes of about equal size, the exopod resting in a groove of the endopod. Endopod is thus gradually being reduced, consisting now of but two segments, and the two segments of the protopod are being gradually broadened and shifted to the inside. There is thus produced the *foliaceous* type of appendage.

The thorax therefore has eight pairs of appendages, the three pairs of maxillipeds and the five pairs of pereiopods, and consists of eight segments.

(3) The head appendages: The head has five pairs of appendages, omitting the eyes. These are, beginning with the most posterior, two pairs of *maxillae*, a pair of *mandibles*, the *antennae*, and the *antennules*.

Examine the *second* maxilla in place. Does it have an epipod? gill? Observe that the epipod is continuous with an anterior process, which is the exopod, the whole forming an elongated blade, pointed at both ends, which fits into the anterior end of the branchial chamber. This blade, the *bailer* or *scaphognathite*, moves back and forth, drawing a current of water through the branchial chamber over the gills from the posterior end of the chamber forward. Within the exopod is a slender endopod, still further reduced and consisting of but one joint, and within this the expanded protopod, with four processes. The second maxilla is thus decidedly foliaceous. Remove it and **draw**.

The *first* maxilla is a reduced foliaceous appendage. Its two inner thin plates are the protopod; the outer slender process is the endopod, exopod being absent.

Remove the first maxilla.   The small process in front of it is not considered to be an appendage but is a part of the lower lip.

The heavy mandible is now exposed.   It consists of a single triangular piece with strong teeth upon its inner edge;   and a small palp, probably the endopod, folded beneath the toothed margin.   Spread the mandibles apart so as to see the mouth opening.   In front of the mouth opening is the cushion-like upper lip, or *labrum*.

Examine the antenna.   In the middle of the ventral surface of its basal segment (coxopod) find the *renal opening*, the opening of the excretory organ. The long many-jointed filament is the endopod;   the thin sharp projection near the base of the filament is the exopod.   **Draw the antenna.**

The antennule or first antenna has a protopod of three joints from which arise two short many-jointed filaments, which are probably exopod and endopod.

This investigation of the appendages shows that there are at least five segments to the head, eight to the thorax, and six to the abdomen, or nineteen appendage-bearing segments.   If the telson is a segment, as seems reasonable, the lobster consists of twenty segments.   Some zoölogists believe that the eyes also represent a segment and raise the number to twenty-one, but considering that eyes occur on unsegmented animals it seems probable that they are not a pair of appendages homologous to the others.

f) The respiratory system and the branchial chamber (Hegner, p. 204): Study the arrangement of the *gills*, or respiratory organs, in the left branchial chamber, where they have already been exposed.   We have noted that one gill is fastened to the epipod of most of the thoracic appendages.   Such gills fastened to appendages are called *podobranchiae*.   Remove the podobranchia and epipod from the third pereiopod, and observe that two more gills are situated beneath it attached to the arthropodial membrane.   These are *arthrobranchiae*.   In the lobster there is still a third set of gills, seen by removing the arthrobranchiae.   Under the two arthrobranchiae note a gill fastened to the wall of the thorax and hence called a *pleurobranchia*.   Each thoracic segment has therefore typically four gills, but not all of them possess the full number, as the student may readily discover.   The crayfish has no pleurobranchiae.

Cut off one of the gills and examine its structure.   It consists of a central axis bearing numerous delicate threadlike filaments.   Examine the cut surface of the axis and note the two canals which it contains, one for blood to enter the filaments, the *afferent* vessel, and one for it to leave the filaments, the *efferent* vessel.   Note the hole left in the thoracic wall where the gill was removed through which the blood vessels pass.

Remove all the gills from the branchial chamber and note that the segmentation of the thorax is now visible.   Examine the region where the extension of the carapace over the branchial chamber was cut off and see that this extension was not really the sides of the thorax (which are concealed by the gills)but merely a downward fold from the median cardiac region.

g) *The special sense organs:* These are best studied before the animal is dissected. The body and appendages of the lobster bear innumerable hairs, many of which are sensory hairs attached to nerve cells and having tactile functions. Many hairs upon the antennules, antennae, and mouth appendages are also organs of taste and general chemical sense. The head appendages are more sensitive than the others.

The eyes of the lobster are compound eyes, i.e., they are made up of a large number of simple eyes, or *ommatidia,* which are radially arranged. The cuticle covering the eye is thin, transparent, and flexible, lacking the calcareous deposit. It is called the *cornea.* Examine it with the hand lens, noting the minute polygonal areas of which it is composed. Each of these is the outer end of one of' the ommatidia. Make a longitudinal section through the eye and eye stalk with a sharp knife and examine the cut surface. Observe the numerous black ommatidia radiating from a central white region, which is the *optic ganglion.* Read Hegner (pp. 205–8) and study Fig. 103. Then remove some of the ommatidia from the section of the eye, tease on a slide in a drop of water, cover, and examine You should be able to identify the *crystalline cone,* the *rhabdome* surrounded by black *pigment cells,* and the nerve fiber leading away from each ommatidium.

The *statocyst,* or *organ of equilibrium,* is a thin sac located in the basal segment of the antennules. Remove an antennule and cut off the ventral wall of the basal segment of the antennule. This reveals a thin-walled sac containing sand grains, attached to the dorsal wall of the segment. The assistant should demonstrate the crescent of *sensory hairs* within the sac (read Hegner, p. 208).

2. **The internal anatomy.**—With a scissors carefully remove piece by piece the whole dorsal half of the body from rostrum to telson, beginning at the anterior end. In doing so the following points are to be observed; hence read the next section before cutting. Be especially careful not to injure the blood vessels, which are injected with a yellow fluid.

a) *Body wall and muscles:* Beneath the hard exoskeleton is a thin membrane, the *epidermis,* which is the ectoderm and secretes the exoskeleton. Various muscles will be found attached to the shell. In the anterior region under the carapace is the large thin-walled *stomach* from the anterior and posterior ends of which *gastric* muscles extend to the carapace. At the sides of the posterior end of the stomach, in front of the cervical grooves, is a heavy fan-shaped mass of muscle. This is the *mandibular* muscle, which moves the mandible. In removing the dorsal exoskeleton of the abdominal segments notice the longitudinal muscles attached to them and trace them forward to their origins on the sides of the thorax. They are the *extensors* of the abdomen, that is, they straighten the abdomen. After the removal of the whole dorsal exoskeleton, and the muscles attached to them, notice the large masses of ventral abdominal muscles, the *flexors* of the abdomen. Why are they more powerful than the extensors? Remove the left thoracic wall, and observe the forward extension of the flexor muscles to their origins from the lateral and ventral thoracic walls.

To illustrate the arrangement of muscles in the appendages, examine those of the chela. Cut off the chela from the first pereiopod and then remove the shell from one surface of it. Find within the two muscles, one much larger than the other, and find by pulling upon them their method of attachment and action on the movable part of the pincer. Pick away the muscle fibers and find the strong tendon in the center of the mass. Each joint of each appendage is similarly provided with an extensor and a flexor muscle for moving the next joint (see Hegner, p. 209 and Fig. 105).

We observe that the muscular system of the lobster is highly developed as compared with that of the lower invertebrates which we have studied. In those forms there exist simple cylindrical tubes of muscle fibers, extending lengthwise, and producing merely extension, contraction, or bending of the body. But in the lobster as in the frog separate definite muscles exist having specific actions on various parts of the body, permitting much greater variety and exactness of movement. They have definite origins and insertions on the exoskeleton. Are the abdominal muscles segmentally arranged?

Place a small piece of muscle on a slide in a drop of water, tease into fibers, cover, and examine. The fibers will be found to be cross-striated, like the voluntary muscles of the frog. On the other hand, the muscles of the lower invertebrates are like the smooth involuntary muscles of the frog.

b) The circulatory system: Your specimen should now have the dorsal and one lateral wall cleaned away so as to expose the viscera. The conspicuous organs are the stomach anteriorly, the *heart* posterior to this, usually injected with yellow material and with yellow vessels springing from it, and a large white organ, the *digestive gland*, occupying the sides of the space within the thorax and extending back into the abdomen.

The space around the heart and between the viscera is not a coelome, as the student might expect, but it is an enormous blood space, or *blood sinus*, filled in life by blood. The coelome has been greatly reduced, in fact, is practically absent through this great development of blood sinuses. The studies which we have made on the coelomes of the earthworm and the frog should convince the student that this space in the lobster cannot be a coelome because mesenteries are completely lacking. Note that none of the organs are supported by mesenteries and that there is no peritoneal lining.

The heart lies free in a large sinus, the *pericardial sinus*. The following arteries arise from it. From the anterior end of the heart in the median line is the single *ophthalmic* artery. Trace it forward along the dorsal surface of the stomach and note that it forks to supply the eyes. On each side of the ophthalmic artery arises an *antennary* artery which curves downward over the digestive gland. Trace it and note its branches to the mandibular muscle, the stomach, the antennae, and the antennules. Directly below the origin of the antennary arteries and from the ventral surface of the heart the paired *hepatic* arteries extend down into the substance of the digestive gland. From the posterior end of the heart a large vessel, the *dorsal abdominal* artery, extends backward the whole length of the abdomen, forking in the sixth abdominal segment. It gives branches to the intestine which is immediately beneath it and to the extensor muscles of the abdomen which have been removed. Are its branches segmentally arranged? From the posterior end of the heart at the same point as the origin of the dorsal abdominal artery, another large artery, the *sternal* artery, arises and proceeds directly ventrally. Push the left digestive gland carefully aside to see it. Its further course will be traced later.

Make an outline of the lobster from the side as large as your drawing page and in this put the heart and its arteries. Other systems will be added to this later.

Remove the heart from the body, wash it in water, and note its peculiar angled form. It is generally somewhat distorted by the injection. Find the three pairs of openings, or *ostia*, in the wall of the heart. One pair is on the dorsal surface, one pair on the ventral surface, and one pair is lateral just under the lateral margins (see Hegner, Fig. 101, facing p. 195).

The circulatory system of the lobster and other arthropods is an *open* system, that is, the only definite vessels are the arteries, and the circulation is completed through open spaces, or sinuses. In the lobster the blood passes from the arteries into these sinuses and finally collects into one large sinus, the *sternal* sinus, in the median ventral line. This will be seen later. It then passes into the gills by the afferent vessels, out by the efferent vessels, already seen in the cross-section of a gill, and back to the pericardial sinus by definite channels. When the heart expands the ostia open and blood is sucked into the heart from the pericardial sinus. The blood is colorless.

The ophthalmic artery, heart, and dorsal abdominal artery together correspond to the dorsal vessel of the earthworm. In the earthworm, it will be recalled, the entire dorsal vessel is contractile, but in the lobster the power of contractility has become limited to one region, which is enlarged and now designated as a heart. There is also in the lobster a vessel to be seen later which corresponds to the ventral vessel of the earthworm. The student should particularly notice that the segmental vessels, connecting these two in each segment of the earthworm, are for the most part lacking in the lobster.

c) The reproductive system: Lying under the heart will be found a pair of slender gonads, white in the male, pinkish in the female. They extend forward almost to the anterior end of the stomach and posteriorly beyond the termination

... the digestive gland in the lobster. (In the crayfish they are shorter and ... the posterior part is single.) The two gonads have a slender con-... in front of the heart. In the female an oviduct will be found arising ... in the region of the heart. Trace it downward over the surface ... digestive gland into the third leg. In the male the vas deferens arises ... in the same region and extends backward to the fifth leg. Soon ... the testis the vas deferens presents a bend, then widens and pro-... to the external opening. (The vas deferens of the crayfish is ... **Draw in the reproductive system** in your lateral view of the lobster.

... **digestive system.** The conspicuous parts of the digestive system are the ... stomach and the voluminous digestive gland. The latter, ... filling the greater part of the cavity of the cephalothorax, ... but as it really combines the functions of both a liver ... would be more appropriately designated the *hepato-pancreas*. ... lobes and small tubules of which it is composed. Note shape, size, ... digestive gland and enter it on your drawing with very light ... the left one completely, noting as you do so the place where ... posterior end of the stomach by an hepatic duct.

... digestive gland exposes the stomach more fully. Observe ... over to the right the short *esophagus* connecting it with ... which is divided by a constriction into a large anterior *cardiac* ... contains hard *ossicles*, and a much smaller posterior *pyloric* ... ducts open into the pyloric portion just above the rounded ... downward from its sides. Fastened to each side of the ... the stomach will often be found a large mass of crystals, ... called a *gastrolith*. See Hegner (p. 201) for its pos-... the end of the pyloric chamber find the slender intestine ... It makes a deep ventral bend just behind the stomach ... a position directly beneath the dorsal abdominal artery. In ... it gives off a blind dorsal sac, or *caecum*; from this ... is called *caecum*. **Draw in these parts** of the diges-... an outline of the lobster.

... leaving the esophagus in place. Cut it open, wash it ... Note the hard pieces, or *ossicles*, in the walls ... the paired ... and single wall or teeth, and the pro-... with the silky hairs on the posterior division of the ... The teeth and ossicles are a ... by the *gastric muscles* ... the hairy processes. ... passing into the

e) The excretory system: In the anterior end of the cephalic cavity, inside of the base of the antenna, locate a circular greenish mass, liberally supplied by branches of the antennary artery. This is the excretory organ, commonly called "green gland." It is in reality a modified nephridium, which has lost its funnel-shaped opening into the coelome (since there is no coelome). By carefully spreading it out with a forceps, determine that it is a blind sac, curved upon itself into a circle. The renal opening in the basal joint of the antenna has already been observed. The student should note that while in the annelids there is a pair of such nephridia in practically every segment, in the lobster they are present in one segment only, the antennary segment. This is another example of the loss of segmental structures, which characterizes the arthropods as compared with the annelids. **Draw** in the green gland on your side view.

f) The nervous system: Remove all soft parts from the interior, leaving the esophagus and sternal artery in place. First begin in the abdomen and carefully remove all of the ventral abdominal muscles, noting their segmental arrangement and the complex manner in which they loop over one another. In the median ventral line of the abdomen, next to the inner surface of the shell, is a white cord, the ventral nerve cord. In the thoracic region this will be found to disappear into a cavity which is roofed over by hard plates. Clean out the muscles and other soft parts in the thorax until you have exposed these plates. They form an internal skeleton, called the *endophragmal* skeleton; this is really produced by ingrowth from the exoskeleton.

The cavity underneath the endophragmal skeleton is the *sternal sinus,* which was mentioned in connection with the circulatory system as the large sinus in which all of the venous blood collects before going to the gills. In this sinus are also located the nerve cord and the branches of the sternal artery. Remove the endophragmal skeleton by cutting along each side and taking out the middle piece, and trace the nerve cord forward in the sternal sinus to the esophagus. Here note that it forks, passing on each side of the esophagus and uniting again into a bilobed mass just within the region occupied by the eyes.

The bilobed mass beneath the eyes is the so-called *brain,* better designated as the *supra-esophageal ganglia.* It consists of two ganglionic masses fused medially, and sends nerves to the eyes (where they expand into the optic ganglia, situated in the eye stalks), the antennules, and the antennae. The brain, therefore, represents at least three pairs of ganglia fused together. From the brain arise the two *circum-esophageal commissures.* Trace these around the esophagus to the large *sub-esophageal* ganglion, just behind the esophagus. Note branches arising from the circum-esophageal commissures and sub-esophageal ganglion. Trace the nerve cord posteriorly along the floor of the thorax, noting the enlargements or ganglia. Behind the sub-esophageal ganglion there are three ganglia; then the nerve cord forks to allow the sternal artery to pass through; behind the sternal artery are two more thoracic ganglia. Since in a segmented animal there is a ganglion, really a pair, to each segment, how many ganglia should there be in the oesophagus of the lobster? Since the brain supplies the first two pairs of appendages how many appendages must be supplied by the sub-esophageal ganglion and of how many fused ganglia does it, therefore, consist?

Trace the nerve cord back into the abdomen. Is there a ganglion for each segment? Does the telson have a ganglion? Note the branches of the abdominal nerve cord.

Observe the branches of the sternal artery under the ventral nerve cord. It passes between the two halves of the nerve cord, between the fourth and fifth thoracic ganglia, and promptly divides into an anterior horizontal branch, the *ventral thoracic artery*, and a posterior branch, the *ventral abdominal* artery. These correspond to the ventral vessel of the earthworm.

Put in the nerve cord and its ganglia accurately, and the branches of the sternal artery on your drawing. The drawing is now complete.

The nervous system of the lobster is very much like that of the earthworm, except that as in the case of the other systems of the lobster it shows a partial loss of the segmental arrangement through fusion. The nervous system of both earthworm and lobster is based upon the same plan as that of *Planaria*, consisting fundamentally of two ventral ganglionated cords arising from a dorsal brain. This type of nervous system is common to all the invertebrates (except those having radial symmetry) and is called the *ladder type*. The ventral cords, originally separated rather widely, come together in the segmented animals to produce an apparently single cord.

General considerations on the lobster. The chief principle which we set to be bringing out through the study of the anatomy of the lobster has already been emphasized throughout the laboratory instructions. It is that the lobster, although a segmented animal like the annelids, exhibits a marked tendency to specialization of the segmentation, through fusion and loss of segments and of appendages functions. This tendency is still further in evidence in the verte-brates, where the section done reveals that the frog is segmented, and then in only a few regions. What systems of the lobster are segmented? Which is the most completely segmented? In what part of the body is the segmentation ... Where is it most obscure? Compare with the frog and deter-mine the systems and the parts of the body which exhibit the greatest ... in structure in the cases of the two animals. Is it the same system ... to be ... which most clearly segmented? What is the signifi-cance ... systems which are present ... segment ... Why is ... Are the systems which the ... more specialized ... worm? Do the ... is to specialization for particular functions? What system of the lobster has made the least progress? Do both the lobster and the frog seemed to have attained a fairly high degree of adjustment (adaptation) to the conditions under which each lives?

The grasshopper is selected as a representative of the great group of insects because it is relatively large, easily obtained, and a rather generalized form. The following description is based upon the large Florida grasshoppers.

1. **External anatomy.**—Obtain a preserved specimen. Compare with the earthworm and especially with the lobster. Is the animal bilaterally symmetrical? Is it segmented throughout? In what part of the body is the segmentation most apparent? least apparent? Does it have jointed appendages like the lobster? Do they occur on every segment? What part of the body lacks them? Is this in accordance with the principle of specialization of anterior regions, which we observed in the other animals studied? Does the animal have a definite *color pattern?*

The body is covered by a *chitinous exoskeleton*, similar to that of the earthworm and the lobster. It is secreted by the ectoderm, or epidermis, which lies beneath it. As in the lobster it consists of hard regions, or *sclerites*, joined together by thin membranes, the *arthropodial membranes*, at lines known as *sutures*.

The body is divided into head, thorax, and abdomen. Are these regions more distinctly separated than in the lobster? Each part consists of a definite number of segments. Each segment as in the lobster is typically composed of a dorsal sclerite, the *tergum* (commonly called *notum* in insects), a lateral sclerite, the *pleuron*, and a ventral sclerite, the *sternum*.

a) The head and its appendages: Is the head readily movable upon the body? The head shows no segmentation. The larger part of it is inclosed in one hard sclerite, the *epicranium*, in which may be distinguished a top (*vertex*), sides (*genae*), and a front (*frons*). Looking at the head from in front, the lower limit of the frons is marked by a distinct transverse suture. Below this suture is a rectangular sclerite, the *clypeus*, and below this and attached to it another sclerite, the bilobed upper lip or *labrum*. Observe that the labrum is movable; it is not, however, an appendage.

The head bears eyes, antennae, and three pairs of mouth parts. There is a pair of large compound eyes situated in the upper parts of the genae. Examine them with the hand lens and note the minute hexagonal areas into which the surface is divided. Each of these is the outer end of an ommatidium, and the structure of the compound eye is the same as in the lobster. The grasshopper also has three simple eyes, or *ocelli*, one anterior to the dorsal portion of each compound eye on the ridge which separates frons and gena, and the third in the depression between the two ridges in the median line of the frons. For the structure of the ocelli see Hegner (p. 245). **Draw a front view of the head.**

The *antennae* are the first pair of head appendages. They are jointed filamentous structures, springing from depressions in the upper part of the frons. The antennae are very important olfactory and tactile appendages. Remove an antenna and examine with the low power, noting the sensory hairs upon it.

With the forceps, remove the clypeus and labrum in one piece, and examine the under surface. This forms the roof of the mouth and bears a central club-shaped elevation, the *epipharynx*, probably of sensory function.

The removal of the labrum exposes the three pairs of mouth parts. The first of these, constituting the second pair of head appendages, is the *mandibles*, very hard brown organs with toothed inner edges. Remove these and **draw one** under a hand lens. The next pair of mouth parts, the third pair of head appendages, are the *maxillae* (first maxillae). They are lateral and each has a conspicuous process, the *palp*. Remove a maxilla complete and study it with the hand lens. It consists of a basal portion, composed of two segments, a lower *cardo* and an upper *stipes*, from which springs three processes. The inner one is a curved, toothed blade, the *lacinia;* the middle one an oval plate, the *galea*, composed of two joints; and the outer, a slender jointed process, the *maxillary palp*, supported by a small basal joint, the *palpifer*. **Draw the maxilla.** It is probable that the palp is the endopod, and the remainder of the appendage the protopod, exopod being absent.

The last pair of mouth parts, the fourth pair of head appendages, is the *labium*, or lower lip, lying below the mouth in the median line. It is composed of two maxillae (the second maxillae) partially fused in the median line. Attached to the labium and projecting inward from it to form the floor of the mouth cavity is an elevation, the *hypopharynx*, which serves as an organ of taste. Remove the labium completely and identify the following parts with the hand lens: the basal crescentic segment, the *submentum;* the next single piece, the *mentum;* the paired *labial palps*, springing from the sides of the mentum through a small joint, the *palpifer;* and the paired flat median plates, the *ligulae.* **Draw the labium.**

Since the head bears four pairs of appendages it must consist of at least four segments. Investigation of insect embryos has shown that there is another segment in front of the one bearing the antennae, and still another between the antennary and mandibular segments. There are thus six segments in the insect head. The antennae of the insects correspond to the antennules of the lobster; while the segment and the appendages which correspond to the antennae of the lobster are lost in the adult insect.

b) The thorax and its appendages: The thorax is composed of three stout segments, called the *prothorax, mesothorax*, and *metathorax*, respectively, beginning at the anterior end. The tergum, or notum, called the *pronotum*, of the prothorax is greatly enlarged and extends back like a shield over the other two segments of the thorax. It is also seen to be composed of four distinct sclerites, one behind the other. The sternum of the prothorax has a sharp posteriorly directed spine; the pleuron is rudimentary. The prothorax bears the first pair of legs. Is the prothorax independently movable?

Cut away the backward extension of the pronotum. In the middle of the side, in the membrane between the prothorax and mesothorax, find an oval opening. This is a *spiracle*, or *stigma*, one of the openings into the respiratory system. In the mesothorax, identify the dorsal tergum, or notum, the lateral pleuron, the ventral broad sternum. Both tergum and pleuron are composed of more than one sclerite. The mesothorax bears dorsal outgrowths, the first pair of wings, and a ventral pair of appendages, the second pair of legs. Cut off the wings.

The metathorax is similar in form and parts to the mesothorax and bears the second pair of wings and third pair of legs. Cut off the wings. Locate another spiracle between the ventral portions of the pleura of the mesothorax and metathorax.

Examine the wings. Compare them as to form, size, color, thickness. What do you consider to be the functions of each pair? The wings arise as saclike outgrowths of the body wall. During development the two walls of the sac become pressed together, forming a thin membrane. The *veins* or *nerves* of the wings are respiratory tubes, filled with air, each surrounded by a tubular blood sinus. After the insect attains its adult size, blood ceases to flow in the wings, and they become dry, hard, and lifeless. Examine a piece of the second pair of wings under the microscope, and note the air tube, or *tracheal* tube, and the blood sinus around it in each of the veins.

Remove a leg from the body noting the depression in the body where it fits, the arthropodial membrane which attaches it to the body, and the muscles at its base. It is composed of five segments. The rounded segment which adjoins the body is the *coxa*. It is succeeded by a quite small joint, the *trochanter*. The next segment is the long powerful *femur*. Beyond this comes the more slender, spiny *tibia*. The terminal part of the leg is the *tarsus*, consisting of three joints, and with five ventral pads, the *pulvilli*, and a terminal pair of hooks. Compare the three legs of the grasshopper with each other. Do all have the same parts? Can you associate differences in relative proportions of parts with differences in function? Are the legs well "adapted" for their functions? What are the uses of the pulvilli and terminal hooks?

c) *The abdomen and its appendages:* Each abdominal segment consists of a tergum, a U-shaped piece forming the dorsal and lateral walls, and a sternum, the convex ventral plate. The pleuron is practically lacking but is represented by the membranous fold where tergum and sternum articulate. Just above this line of junction, on the lower border of the tergum will be found a spiracle, or

... spines.

... Behind the ... which ... The *female* ... There ... minal appendages. ... lays its eggs in ... the styles?

... specimens is rather ... difficulty in locating ... searching for

... opening in front of the suranal ... Remove and preserve ... the viscera is revealed. ... is an enormous blood sinus; this is evidenced by the complete lack of mesenteries. The coelome is in fact practically entirely wanting in insects. A yellowish material, the *fat body*, representing stored food, will be found attached to the viscera and body wall. Note also the slender bandlike muscles originating on the body wall and attached to the movable appendages.

Place the specimen in a wax-bottomed dissecting pan, pin it down by pins through the wall, and fill the pan with water.

a) *The respiratory system:* There is perhaps no other system to be found in the animal kingdom which so excites the admiration of the zoölogist as the respiratory system of the insects. Unfortunately it is impossible to study it satisfactorily in preserved material. It consists of a series of tubes, symmetrically and segmentally arranged, called the *tracheal tubes*, or *tracheae*. These open to the exterior through the spiracles, or stigmata, which have already been noted on the surface of the body. To see the tracheal tubes, push the viscera of the grasshopper gently to one side and look on the inside of the body wall opposite the points where the spiracles are situated. A white tube will be seen extending inward from each spiracle. With a little practice one will soon be able to identify similar tubes throughout the body. Note also particularly in the thorax the large white *air sacs*, which are connected with the tracheae and serve to increase the capacity of the tracheal system. The tracheal system of the grasshopper consists of three pairs of longitudinal trunks, connected by segmental branches, giving off branches to all parts of the body and provided in certain places with air sacs. Recall the tracheal system seen in the fly larvae (Section II, F, 3), and examine Hegner's Figs. 135 (p. 243) and 137 (p. 244).

Remove a piece of a tracheal tube, mount in a drop of water on a slide, and examine with the microscope. Observe the spiral thread on the inside of the tracheal tube. What is its function? **Draw.**

The tracheae are tubes produced by ingrowth of the ectoderm and lined therefore by the same chitinous layer which covers the body, in which the spiral threads arise by local thickenings. The tracheae arise therefore in a manner opposite to that of gills, which are outgrowths of the body wall. Both serve the same purpose, an increase of surface. But while gills contain blood which obtains oxygen through their thin walls, the tracheal tubes are filled with air. The finest ramifications of the tracheal tubes are in contact with single cells; and further, the blood in the large blood sinuses is in contact with the walls of the tracheae. Thus oxygen is brought directly in contact with all parts of the body, even the smallest; and a respiratory system which is regarded as the most efficient among animals results. The air in the tracheal system is changed through respiratory movements of the body.

b) *The circulatory system:* Associated with the development of a remarkably efficient respiratory system, there is a correspondingly poor differentiation of the circulatory system. There is a long segmented *heart* situated in the median dorsal line of the abdomen. Look on the underside of the dorsal strip removed from the animal for a long tubular structure with several dilatations. It is often indistinguishable in specimens which have been preserved for a long time. Fan-shaped muscles are attached on either side to each dilatarium; they are called the *alary muscles* and are supposed to help dilate the heart. The wall of the heart is pierced with *ostia*, as in the lobster, through which it sucks in blood from the great blood sinus which surrounds it.

There are no arteries or other blood vessels, but the blood flows out from the anterior open end of the heart into the sinuses.

c. The reproductive system: This system will be considered first since it is the first system noticeable on looking into the cavity of the abdomen. In the male the testis forms a white mass in the posterior end of the animal dorsal to the intestine. From each side of the ventral surface of this mass a male duct or *vas deferens* arises and passes posteriorly and ventrally to the end of the abdomen where it forms a coiled mass, located in the seventh and eighth abdominal segments. The lateral walls of these segments should be removed to see these masses. The two vasa deferentia then unite in the ventral median line within the eighth sternum to a single duct which passes into a muscular mass surrounding the base of the *copulatory organ,* or *penis.* Remove the ninth sternum and the subgenital plate and look for these parts. Remove the muscles from the penis and note with a hand lens or microscope the four hard chitinous styles which compose it.

In the female the two *ovaries* are situated in the posterior end of the abdomen, one on each side of the digestive tract. Each consists of a number of parallel *ovarian tubes,* more or less vertical in position, resting, so to speak, in a row upon the oviduct which arises from their ventral ends. In each ovarian tube is a row of eggs of which the largest ones, often plainly visible as oval brown bodies, are placed nearest the beginning of the oviduct. The two oviducts pass posteriorly and centrally and unite in the median ventral line under the intestine to a common duct, the *vagina.* The vagina opens to the exterior by the female genital opening located at the base of the egg guide. Grasp the egg guide with a forceps and pull it forcibly out of its position between the ventral styles of the ovipositor. At its base note a little cushion which folds over the genital opening and must be pulled back to reveal the latter. Remove the eighth sternum and locate at the end of the vagina a broad tube passing to the genital opening. Lift out the vagina and note just above it another tube, the *copulatory sac,* which receives the sperm from the male in copulation. This opens to the exterior in common with the vagina. Attached to the copulatory sac is a slender duct leading from a little sac placed just above the point where the two oviducts unite to form the common duct, the *common oviduct.* In laying eggs the grasshopper digs a hole in the ground with the ovipositor; the eggs are then passed out through the genital opening, receiving sperm from the copulatory sac and being stuck together into packets by a fluid from the cement gland. The egg guide directs the eggs in their passage to the exterior.

Draw from the side the parts of the reproductive system which you have been able to find.

d) The digestive system: The larger part of the internal cavity is filled with the digestive tract. Remove the reproductive system. Clean away the fat body and cut off one lateral side of the abdomen and thorax and the same side of the head. Identify the short *esophagus* extending inward from the mouth and opening into a large *crop* which fills most of the cavity of the thorax. Behind the crop is the elongated *stomach*, or *ventriculus*, partially concealed by six thin-walled *gastric caecae*. Each gastric caecum is attached by its middle portion to the wall of the stomach so that one of its pointed ends projects forward and the other backward from the point of attachment. The anterior part of the stomach, covered by the caecae is distinguished as the *proventriculus* or *gizzard*, but it is not well differentiated in the grasshopper. The posterior end of the stomach is marked by the presence of a tangle of threads, the *Malpighian tubules*. Beyond this region is the *intestine*, at first wide, then presenting a short narrow portion, the *colon*, then widening into a *rectum*, which extends to the anus. The rectum frequently contains a cylindrical pellet of faeces. Its surface is marked off by longitudinal muscle bands into six expanded areas, known as the *rectal glands*. The position of the anus under the suranal plate has already been noted. Look in the sides of the thorax among the muscles and fat body for the *salivary glands*, a cluster of small round glands attached to a duct. The ducts open into the mouth.

Draw an outline of the animal from the side and put in the parts of the digestive tract. Other systems which the student has seen well enough may also be entered upon the drawing.

e) The excretory system: The threadlike Malpighian tubules arising at the junction of the stomach and intestine are the excretory organs. They are outgrowths of the digestive tract and are therefore entirely distinct morphologically from the excretory organs of the other animals we have studied, that is, the nephridia. The insects have lost the nephridia.

f) The nervous system: Cut through the esophagus, leaving it in place, and remove the digestive tract. Look in the median ventral line of the abdomen for the *ventral nerve cord*. How many ganglia are present in the abdomen? Trace the nerve cord forward into the thorax. Note here, as in the lobster, a so-called endoskeleton, really an ingrowth of the exoskeleton, extending like beams across the floor of the thorax. Remove these and muscles, etc., so as to expose the thoracic nerve cord. Find the three large thoracic ganglia, one in each segment of the thorax. Anterior to these is the *sub-esophageal* ganglion. From this the *circum-esophageal commissures* pass forward around the esophagus, and unite at the brain, or *supra-esophageal* ganglion, located between the eyes.   Does the nervous system of the grasshopper resemble that of the lobster?

Draw in the nervous system on your lateral view of the grasshopper.

g) *The sense organs:* The chief sense organs have already been noted.   The following additional statements may be made.   Read also Hegner (pp. 245–48). Insects are provided with a great variety and abundance of sense organs.   Much work still remains to be done upon the structure and functions of these.   Organs of touch in the form of tactile hairs are present all over the body, but particularly on the antennae, mouth parts, and cerci.   Organs of taste are present as sensory pits on the epipharynx, hypopharynx, and probably on the maxillary and labial palps.   The sense of smell is incredibly keen in insects, and is located mainly upon the antennae in the form of sensory pits; it is probable that olfactory organs are present elsewhere also.   Hearing is localized in the grasshopper in the chordotonal organ; in other insects hearing may be served by auditory hairs and auditory pits.   The compound and simple eyes are the organs of vision; there is no doubt that insects see objects, and they may even perceive colors, but they are especially quick in detecting movements of external objects, a capacity undoubtedly due to the compound structure of the eyes.

3. **General considerations on the grasshopper.**—What are the chief differences between the external anatomy of the grasshopper and the lobster?   What internal systems of the two animals show the most differences?   What systems of the grasshopper are segmented?   What particularly effective and highly specialized systems does the grasshopper possess?   From your study of the anatomy of the grasshopper, can you give reasons why insects are the most successful and dominant animals on the earth, excepting man?

C. GENERAL SURVEY OF ARTHROPODS

Examine the specimens and compare them with each other and the lobster and grasshopper as to degree of segmentation, specialization and number of appendages, general form, symmetry, and divisions of the body.   The common groups of arthropods are:

1. **Crustacea,** forms with two pairs of antennae, many pairs of appendages, hard crustaceous exoskeleton, and gills as respiratory organs.   Examine the following representatives (not arranged by classification).

a) *Entomostraca,* small to microscopic Crustacea, common in fresh water, often inclosed in a bivalve carapace.   Preserved forms are of little value; living forms if available will be demonstrated.

b) *Barnacles,* sessile forms, inclosed in calcareous plates.   Marine forms, covering the rocks along the seashores.

c) *Shrimps*, marine animals similar to crayfishes, but with a more slender, delicate build.

d) *Crabs*, marine animals like the lobster, but with a much broadened cephalothorax, and reduced abdomen folded under the cephalothorax.

e) *Amphipods*, small fresh-water animals, something like miniature crayfishes, but strongly compressed from side to side.

f) *Sow bugs* or *pill bugs*, small forms, greatly compressed dorsoventrally, common in damp places, greenhouses, etc., also in water; curling into balls when disturbed.

2. Arachnids, forms without antennae, and four pairs of walking legs.

a) *Spiders*, require no description.

b) *Scorpions*, with segmented abdomens narrowing to a long tail, bearing a terminal sting.

c) *Daddy longlegs*, or *harvestmen*, with small bodies, indistinctly segmented, and very long slender legs.

d) *Mites*, small, flattened forms, having lost all traces of segmentation, generally external parasites.

3. The king crab, or horseshoe crab, survivor of an ancient, mostly extinct, group, with a large rounded cephalothorax; abdomen terminated by a long sharp spine.

4. Insects, with one pair of antennae, three pairs of walking legs, breathing by means of tracheae.

5. Millipedes and centipedes, elongated forms consisting of many segments. Each segment has one pair of legs in the centipedes, and two pairs in the millipedes.

1. COMPARISON OF CROSS-SECTIONS

Make diagrammatic cross-sections through *Hydra. Planaria*, earthworm, and f. g. cutting in all the layers of the body. Color ectoderm blue, mesoderm red, entoderm yellow, peritoneum and mesenteries green. Indicate epithelial layers by outlines at right angles to the surface: connective tissue by solid shading, muscles by diagonal lines, skeleton by dots. After finishing these diagrams, may them ask be sure that you understand thoroughly what layers correspond in the four animals, and in what important ways the sections differ from each other.

2. COMPARISON OF FUNCTIONAL SYSTEMS

Make a table as follows. Rule off a page with vertical and horizontal lines so as to make a number of vertical and horizontal columns. At the left, in the spaces between the horizontal lines, write the names of all the animals studied in this course, beginning with the amoeba and ending with the frog, in their proper invertebrate order. At the tops of the vertical columns, going from left to right, write the names of the systems in the following order: digestive system, muscular system, nervous system, reproductive system, excretory system, circulatory system, respiratory system, skeletal system. In the appropriate spaces opposite the names of the animals, write in very briefly whether it has such a system and if so what it consists of. The table will show at a glance the gradual increase in numbers of systems as we pass upward from system through the animal kingdom. Note that a system necessary for a animal scale for the performance of specific functions. A snake reveals, a amoeba of scale for the performance of specific functions. Thus the amoeba has and it has no respiratory system, that is, no structures of the surface.

XV. EXERCISE ON CLASSIFICATION

The science of *classification* or *taxonomy* is that branch of biology which attempts to discover the natural relationships of organisms, and to put all known forms in life in their proper places in a genealogical tree. Classification is based entirely upon anatomy, both adult and embryonic, and is not in the least concerned with function.

Classification starts with the conception of the *species*, which may be defined for our purpose as a group of organisms essentially alike in all the details of their structure. Thus all of the frogs given out in the laboratory are so nearly alike that the laboratory instructions even down to the microscopic structure apply to all individuals. Such frogs therefore constitute a species and are given a name, called the *specific name*, which is in this case *pipiens*. There are many other kinds of frogs which are quite similar to this one but differ in small details; for instance, the bullfrog is larger, has a different color pattern, and lacks the dermal plicae; it therefore receives another specific name, *catesbiana* (see Holmes, pp. 18–21). There are in fact about one hundred and forty different kinds of frogs, each of which has a specific name. In order to express the fact that all of these frogs are very similar to each other they are placed together into one group, called a *genus*. This genus to which the frogs belong is called *Rana*, from the Latin word meaning "frog," and this name is spoken of as the *generic name*. The full name of our common frog is therefore *Rana pipiens*. This system of naming animals with two names is called the *binomial system* of nomenclature and was devised by Linnaeus, one of the early biologists who became interested in classification (see Hegner, p. 270).

Among the members of a species there are often minor variations, which are distinguished as *varieties*, when sufficiently important and common. Varieties occur most frequently in domestic animals. See, for instance, Hegner's Fig. 160 (p. 294) for a photograph of the varieties of the domestic pigeon.

All of those genera which are naturally related to each other as shown by their structure are united together into a *family*. Thus the lions, tigers, leopards, lynxes, and other catlike animals form a natural family, the *Felidae*, or cats, having sharp fangs, and retractile claws. Similarly the dogs, wolves, foxes, etc., form another natural family, the *Canidae*, in which the claws are not retractile; the bears are another family; so also the hyenas, and the otters, weasels, and martens. All of these families, together with many others which have not been mentioned, have certain characters in common, such as that they all are carnivorous and have strong fangs and sharp cutting teeth, that they have claws, and

127

walk upon their tiptoes, so to speak, and that their skeletons are very similar. In recognition of these facts, which demonstrate that they are related to each other, they are united into a larger group, called an *order*, in this case, the order *Carnivora*. If we consider other animals familiar to us we find that they too fall into natural orders; thus the cattle, sheep, deer, horses, resemble each other in that they walk on their toenails, which are broadened into hoofs, their legs are much modified for running, and their teeth are broad and ridged for grinding up vegetable food. They constitute another order, the *Ungulata*. Similarly the rabbits, squirrels, mice, rats, beavers constitute the order *Rodentia*, distinguished by sharp, chisel-like front teeth.

If we further consider these and other orders of familiar quadruped animals we find that they have certain large characters in common, such as that they are all clothed with fur, that their young are born alive and nourished with milk, and that their skeletons are quite similar. These facts point to undoubted relationships between them, so that they are united into a still larger taxonomic group, the *class*, in this case the class *Mammalia*, or the *mammals*.

The birds are another natural class, distinguished by their covering of feathers, modification of the fore limbs into wings, and egg-laying habit. Frogs, toads, and salamanders are another, with slimy smooth skins; snakes, lizards, alligators and similar animals form another class, with dry scaly skins; fishes constitute a class distinguished by the presence of gills and fins.

All of these apparently diverse classes of animals have further certain common characters, such as the presence of an internal cartilaginous or bony skeleton, consisting of skull, vertebral column, limb girdles, etc., two pairs of appendages, a ventral chambered heart, a dorsal nervous system, etc. They therefore together constitute one of the great divisions of the animal kingdom, a *phylum*.

The taxonomic divisions are therefore variety, species, genus, family, order, class, phylum. There are usually other subdivisions, also, as subphylum, subclass, suborder, superfamily, subfamily, etc. Naturally, the details of classification are not yet agreed upon because we know as yet little about the natural relationships of animals and because it is difficult to decide whether certain characters are as important as others, as, for example, whether certain differences between two animals will place them in different genera only, or whether they are great enough to separate them into two different families. For this reason the student need not be surprised to find that the various textbooks do not agree on the details of classification, although all recognize the same large groups.

There follows a key to the phyla and classes of the animal kingdom. Peculiar, rare, and aberrant forms are not provided for in the key, but an attempt has been made to include all animals commonly met with. **With the aid of the key,** classify ten different animals provided by the assistant. In this key each statement of characters bears a number followed in parentheses by one or more numbers which refer to the alternative statement. If, therefore, upon reading the first statement the student decides that this does not fit the animal in question, he turns at once to the number or numbers in parentheses and continues to do this until he finds a statement that does fit. In this case he proceeds to the number given at the end of the statement. This method of making a key is taken from Ward and Whipple's *Fresh-Water Biology.* The key is in part derived from one devised by Dr. V. E. Shelford, of the University of Illinois. Only the simplest possible characters have been chosen as means of identification, even though this often involves repetition and increased length of the key.

Key to the Phyla of Animals

1 (2). Animals consisting of a single cell, or of a colony of like cells, or masses of multinucleate protoplasm; mostly microscopic. **Phylum Protozoa**

2 (1). Animals consisting of many cells, of several or many different kinds, arranged in definite layers. 3

3 (4, 11). Without definite symmetry. Forming sessile motionless crusts or masses, often branching; body porous, rough, and bristly, pierced by numerous holes, of which one or more are large and conspicuous. **Phylum Porifera**

4 (3, 11). With definite radial symmetry. 5

5 (10). Relatively simple animals, without anus, coelome, or definite organs. 6

6 (7). Sessile, vase-shaped, or cylindrical forms, porous, rough, bristly, with one large terminal, non-closeable opening; without tentacles. A few members of the **Phylum Porifera**

7 (6). Soft, often gelatinous animals, not porous or bristly; apical opening a closable mouth; nearly always with tentacles. Parts of the body arranged in fours or sixes or indefinite. 8

8 (9). With eight radial rows of ciliated swimming plates; tentacles, if present, without nematocysts. **Phylum Ctenophora**

9 (8). Without such rows of ciliated plates; with tentacles armed with nematocysts; often sessile and colonial, some free-swimming.
Phylum Coelenterata

10 (5). More complex animals, with anus, coelome, and definite organs; parts of the body almost always in fives, sometimes indefinite; hard, spiny, or leathery animals; tentacles, if present, branched, and never with nematocysts.
Phylum Echinodermata

11 (4, 3). With definite bilateral symmetry, at least in part of the body; sometimes spirally coiled in part; sometimes posterior end bent anteriorly toward mouth. 12

12 (34, 37). Without an internal cartilaginous or bony skeleton in the form of skull or vertebral column, wall of the pharynx not pierced with gill slits. 13

13 (14). With one or more girdles or crowns of cilia borne upon the discoidal, platelike, or lobed anterior end; with internal movable chitinous jaws; small or microscopic aquatic forms.   **Phylum Trochelminthes**

14 (13). Without such crowns of cilia.   **15**

15 (27). Body not divided into segments, nor with segmentally arranged bristles, nor with segmented appendages.   **16**

16 (17, 20). Wormlike animals without an anus; flattened dorsoventrally, often leaflike or ribbon-like; free-living in water, or parasitic.
                                                        Phylum Platyhelminthes

17 (16, 20). Wormlike animals with an anus; without an exoskeleton; heads poorly developed and not distinct from body.   **18**

18 (19). With a long proboscis inclosed in a sheath lying dorsal to the alimentary canal, capable of protrusion from the anterior end; slender, often very long, flattened to cylindrical worms, mostly marine.   **Phylum Nemertinea**

19 (18). Without such a proboscis; slender cylindrical worms, pointed at each end, covered by a smooth thick cuticle; water, damp earth, or parasitic.
                                                        Phylum Nemathelminthes

20 (17, 16). Generally not wormlike, with an anus; often provided with an exoskeleton, consisting of calcareous shells of one or more pieces, or calcareous or horny cases or tubes, or a gelatinous secretion; if naked, with much more definite heads than under 16 and 17.   **21**

21 (22). Colonial animals, sessile, fastened in cases, tubes, or on the surface of gelatinous masses; provided with a circular or horseshoeshaped crown of ciliated tentacles; small, usually microscopic.
                               Phylum Molluscoidea; Class Bryozoa (Polyzoa)

22 (21). Not colonial, nor microscopic, exoskeleton in the form of a calcareous shell of one or more pieces, or absent.   **23**

23 (24). With definite heads, sometimes bearing tentacles armed with suckers; shell, if present, not bivalve.   **Phylum Mollusca**

24 (23.) Without definite heads; shell bivalve.   **25**

25 (26). Halves of the shell dorsal and ventral in position; nearly always fastened by a stalk protruding between the valves at their hinge.
                                Phylum Molluscoidea; Class Brachiopoda

26 (25). Halves of the shell lateral; never with such a stalk, although sometimes sessile.   **Phylum Mollusca**

27 (15). Body divided into segments, or with segmentally arranged bristles or internal organs, or with segmented appendages.   **28**

28 (33). Without jointed appendages; wormlike.   **29**

29 (30). Remarkably elongated, flat, tapelike worms, without mouth, anus, or digestive tract; head and anterior segments very small; segments very numerous, increasing greatly in size posteriorly, where they drop off when ripe; always internal parasites.   **Phylum Platyhelminthes**

30 (29). Not tapelike; with anus, mouth, and digestive tract; segments decreasing in size posteriorly or of the same size throughout.   **31**

31 (32). Segments numerous, usually decreasing in size posteriorly; without tracheae, tracheal gills, or spiracles; never with definitely differentiated chitinized heads; often with segmental bristles, but these never in large tufts at one end of the body.   **Phylum Annelida**

32 (31). Segments few in number, not exceeding eleven or twelve; not markedly decreasing in size posteriorly; with tracheal tubes, tracheal gills, or spiracles; often separate hard chitinized heads; and may have bristles in tufts at one end.   **Phylum Arthropoda (insect larvae)**

33 (28). With jointed appendages, at least on the anterior segments of the body.   **Phylum Arthropoda**

34 (12, 37). Wall of the pharynx pierced with gill slits; without an internal cartilaginous or bony skeleton.   **35**

35 (36). Small, fishlike forms; sides of the body marked with zigzag muscle segments; mouth without jaws, in the center of a rounded hood.

Phylum Chordata; Subphylum Cephalochorda

36 (35). Inclosed in a saclike gelatinous or tough coat, which has two openings to the outside, for the ingress and egress of water; sessile, solitary or colonial (the latter quite small), or free-floating, marine.

Phylum Chordata; Subphylum Tunicata

37. With a cartilaginous or bony skull and vertebral column.

Phylum Chordata; Subphylum Vertebrata

Key to the Classes of the Principal Phyla

I. Phylum Protozoa

1 (9). With temporary or permanent extensions of the surface of the body, mostly for locomotor purposes.   **2**

2 (3, 6). Extensions in the form of changing pseudopodia, blunt to long and threadlike.   **Class Sarcodina (Rhizopoda)**

3 (2, 6). Extensions as raylike, non-motile (or nearly so) projections.   **4**

4 (5). Rays with a terminal knob.   **Class Suctoria**

5 (4). Rays pointed, without a knob.   **Class Sarcodina**

6 (2, 3). Extensions in the form of long or short, hairlike, very active processes.   **7**

7 (8). Hairlike processes short and numerous (cilia).   **Class Ciliata**

8 (7). Hairlike processes long and few (one or two to several).

Class Flagellata

9 (1). Without locomotor or other processes of the body in the adult state; parasitic.   **Class Sporozoa**

II. Phylum Coelenterata

1 (4, 7). Animals of the hydroid type; sessile.   2

2 (3). Nearly always colonial; body·of the hydroids on the end of slender stalks; without an esophagus, and gastrovascular cavity, a simple sac, not divided by septa.   **Class Hydrozoa**

3 (2). Solitary or colonial; cylindrical or columnar, not divided into body and stalk; with an esophagus, and gastrovascular cavity divided into compartments by partitions.   **Class Anthozoa**

4 (1, 7). Animals of the medusa type, solitary, free-swimming.   5

5 (6). With a velum and of simple structure.   **Class Hydrozoa**

6 (5). Without a velum, and more complex in structure, usually with highly branching gastrovascular canals.   **Class Scyphozoa**

7 (1, 4). Complex floating colonies containing both hydroid and medusa types of individuals.   **Class Hydrozoa; Order Siphonophora**

III. Phylum Platyhelminthes

1 (4). With mouth and digestive tract; not especially elongated, nor divided into segments.   2

2 (3). Free-living, ciliated forms, without suckers.   **Class Turbellaria**

3 (2). Parasitic, not ciliated; with at least one sucker, often more, often hooks in addition.   **Class Trematoda**

4 (1). Without mouth or digestive tract; nearly always very long and tapelike, and divided into segments (not true segments); parasites.
Class Cestoda

IV. Phylum Annelida

1 (4). Segmental bristles present; indefinite number of segments; no suckers.   **Class Chaetopoda** 2

2 (3). Bristles numerous, generally on lateral outgrowths, the parapodia.
Subclass Polychaeta

3 (2). Bristles few, set directly into the body wall; no parapodia.
Subclass Oligochaeta

4 (1). Segmental bristles absent; limited number of segments; with an anal and an oral sucker.   **Class Hirudinea**

V. Phylum Echinodermata

1 (8). With a well-developed skeleton of calcareous ossicles or plates; usually spiny; radial; no tentacles.   2

2 (3). Mostly sessile, attached by a stalk springing from the aboral surface; if free, moving on the aboral surface.   **Class Crinoidea**

3 (2). Not sessile, nor stalked; moving on the oral surface.   4
4 (7). Star-shaped, five to many rays.   5
5 (6). Rays slender, sharply marked off from the disk.
                                                        Class Ophiuroidea
6 (5). Rays broad, not sharply separated from the disk.   **Class Asteroidea**
7 (4). Not star-shaped; spherical, oval, or flattened disks; very spiny.
                                                        Class Echinoidea
8 (1). Skeleton rudimentary; animals with leathery and tough body walls;
elongated, even wormlike, with a tendency to bilaterality; commonly with
branched tenacles around the mouth.   **Class Holothuroidea**

VI. Phylum Mollusca

1 (2). With arms provided with cuplike suckers; shell apparently absent,
or, if present, spiral and divided into chambers.   **Class Cephalopoda**
2 (1). Without such sucker-bearing arms; shell never divided into chambers,
sometimes absent.   3
3 (4, 5). Shell bivalve.   **Class Pelecypoda**
4 (3, 5). Shell of eight pieces.   **Class Amphineura**
5 (3, 4). Shell univalve or absent.   6
6 (7). Shell shaped like an elephant's tusk.   **Class Scaphopoda**
7 (6). Shell generally spirally coiled, sometimes cap-shaped or conical,
sometimes absent.   **Class Gasteropoda**

VII. Phylum Arthropoda

1 (2, 5). With two pairs of antennae in front of the mouth; generally
covered with a hard exoskeleton; breathing by means of gills; usually with
numerous appendages; aquatic or in damp places.   **Class Crustacea**
2 (1, 5). With one pair of antennae; breathing by means of tracheae.   3
3 (4). Elongated, wormlike, with one or two pairs of jointed walking legs
on each segment of the body.   **Class Myriopoda**
4 (3). With only three pairs of jointed walking legs (larval forms of this
class are provided for in the key to phyla).   **Class Insecta**
5 (1, 2). Without antennae; four pairs of walking legs (in some cases
appearing as six pairs through leglike condition of mouth parts).
                                                        Class Arachnida

VIII. Subphylum Vertebrata

1 (4). Fishlike animals, living in water and breathing by means of gills;
without lungs or limbs.   2
2 (3). Without jaws, paired fins.   **Class Cyclostomata**
3 (2). With jaws and paired fins.   **Class Pisces**

4 (1). Not fishlike; always with lungs (sometimes gills in addition); paired appendages in the form of limbs or absent.   5

5 (6). Skin naked and slimy; never marine.   **Class Amphibia**

6 (5). Skin provided with an exoskeleton consisting of hairs, scales, or feathers (if apparently absent, large marine forms).   7

7 (8, 9). Exoskeleton in the form of scales, never with hair or feathers.
                                                          Class Reptilia

8 (7, 9). Exoskeleton in the form of feathers and scales.   **Class Aves**

9 (7, 8). Exoskeleton in the form of hairs (sometimes apparently absent, sometimes scales present also); nearly always nourishing the young with milk.
                                                         Class Mammalia

XVI. EXERCISE ON ECOLOGY

Ecology is that part of biology which studies living organisms in their natural environments. Its problems are: (1) to locate every species of plant and animal in the place which it naturally inhabits; (2) to find out and measure all of the factors which together make up its surroundings, as physical factors (light, temperature, moisture), chemical (oxygen content, carbon dioxide content, salts present), physiographic and metereological (climate, seasonal changes, composition of soil, etc.), and biological (other organisms present); (3) to discover what structures the animal possesses which enable it to maintain itself successfully in its environment (e.g., if an animal living in a swift stream had not some means of hanging to objects, it would be swept away); (4) to determine how its behavior enables it to continue to live in the environment in which it is found and how it responds to the various stimuli present in that environment, a field of work called *animal behavior;* (5) to study how environments change through physiographic or other processes, and how this effects the organisms inhabitating those environments. All of these matters must be determined not only for the adult but also for all stages of the life-cycle. The problems of ecology are therefore exceedingly complex and difficult ones.

The class should be conducted if possible on an excursion into the field, into any characteristic habitat, as a pond, swift stream, woods, etc. Each student should carry a number of Mason jars into the field and bring back to the laboratory living animals upon which experiments can be performed in the laboratory. The animals are to be placed as soon as possible into conditions imitating the natural environment. As this is easiest in the case of pond animals, the pond environment will prove one of the most suitable for an exercise of this kind. While in the field the student will note carefully the various possibilities for animal habitats in each general environment, as in a pond, bottom of the pond, on vegetation, free in the water, under floating objects, etc., and note which animals live in these various places.

As a sample exercise in ecology the following experiments on the pond snail are presented. The common pond snails are *Limnaea*, having a long spiral shell; *Planorbis*, flat spiral shell; and *Physa*, short spiral shell, with the last chamber much larger than any of the others. The instructor should furnish references for reading in connection with this exercise.

1. **Natural environment of the pond snail.**—Where were the animals found in the field? What were they doing? What are they doing now in the laboratory aquaria where they have been kept since collected? What are the factors of

their environment? How does it change in the various seasons of the year? During the day? Is the top water of a pond different from the bottom as to light, temperature, content of oxygen, carbon dioxide, etc.? What chemical substances do you think are present in pond water? Do the snails occur where there is vegetation? How does vegetation change the water chemically? How would it affect the penetration of sunlight into the pond? What is found on the bottom of the pond and what rôle does vegetation play in its production?

2. **Structures of the pond snail which are related to the environment.**—How does the animal move about? The broad flat surface on which it crawls is the ventral surface of the body and is called the *foot*. Is the movement of the foot a muscular one or not? Can you see muscular waves passing along it? Considering the method of locomotion of the pond snail where would they occur in a pond? Can they swim free in the water? How would they reach the surface of the water? Can the animal go up and down in the water, without climbing on objects? If you see it doing so it is traveling on a *mucus thread* secreted by the foot. Remove a pond snail and throw it forcibly back into the water. Does it sink? Does it let out bubbles of air? The pond snails do in fact possess a chamber filled with air. By changing the amount of air in this chamber, could they rise and sink in the water at will?

The rounded projection in front of the foot is the *head*. It bears a pair of *tentacles* at the bases of which is a pair of eyes. Touch the tentacles. What do you think is their function? On a snail which is crawling along the sides of the vessel so that you can see the underside, observe that a fold of the body, the *mantle*, is fastened to the inside of the shell. In the center of the ventral side of the head is the mouth opening.

Pick up a snail out of the water. How far can it withdraw into the shell? What is the purpose of this reaction? What is therefore the advantage of the shell? What disadvantages does its possession entail? What does a snail do when it is violently disturbed? Is this reaction ample to protect it from enemies?

Observe the mud in the bottom of the jars where snails are kept, or else put some snails in a jar of water with mud in the bottom. Note the trails left by the snails as they crawl. Stir up a trail and observe that the particles of dirt stick together. How does the snail accomplish this? Is this of advantage in crawling over soft mud or slippery surfaces?

3. **Food taking.**—As the animal crawls along on the side of the aquarium observe a wedge-shaped object protruding from the mouth at regular intervals. It is the *radula*. The radula is a horny ribbon covered with teeth, used by the snail for feeding. It is worked by a complicated muscular apparatus back and forth over the end of a hard *radular cartilage*, like a rope over a pulley, and thus exerts a rasping action on the food, reducing it to minute bits which are then sucked into the esophagus. What does a snail get by rasping the glass of the aquarium? Do you see evidence that the snails have been feeding on the plants

in the aquarium? What would they get besides plant tissue by scraping the surface of vegetation? Scrape the surfaces of various aquatic plants and examine the scrapings with a microscope. What do you see (A)? Would you expect to find snails in bodies of water where vegetation is scanty or is composed of harsh, tough plants? This gives us a clue as to where to look for pond snails and as to one of the factors which limit their distribution. Examine a slide of the radula and note the teeth upon its surface.

4. Respiration.—Watch the snails for some time and observe that they eventually come to the surface and assume a characteristic attitude. (How can a heavy animal like a snail keep on the surface?) Observe between the head and the edge of the shell, which is covered by the mantle, a small conical projection. This is the *respiratory* or *pulmonary sac.* Observe that the tip of the sac is thrust above the surface film, and is then opened to the air. The animal remains in this attitude a little while taking air into the sac. It then closes the sac and descends. The common pond snails therefore breathe air, but there are some fresh-water snails which breathe by means of gills and do not have to come to the surface. By changing the amount of air in the pulmonary sac, snails can rise and sink in the water.

What stimulus do you suppose drives the snail to the surface for air? What factors would determine how often it would need to come to the surface? Try difference in rate of taking air between snails kept at ordinary temperatures and those in ice water. How does the snail know in which direction to go for air? To determine this, take a wide-mouthed bottle, put some snails in it, fill it completely with water, and cork tightly so that no air bubbles are included. Turn it upside down and watch in which direction the snails go to seek air. Does this answer the question?

How long will snails live without access to oxygen? Take three bottles of equal size, place an equal number of snails in each and some vegetation for food. Fill two bottles completely with water and stopper tightly. Place one in the ice box, leave the other at room temperature. Fill the third bottle partly with water and leave open in the room. How long do the snails survive in the stoppered bottles? Does temperature make a difference? Why? How would this apply to the living conditions of snails in winter, when the ponds are covered with ice and they cannot get to the surface?

5. Desiccation experiment.—Take a glass jar, put about an inch of water in it, and several snails. Let stand until the water is completely dried up. What do the snails do? Can they carry on activities in the absence of water? Pick one up and examine. What do you find across the mouth of the shell? Are the snails dead? Put some water in the jar and note results. How does this apply to the possibility of ponds drying up in hot weather?

6. Reaction to light.—Prepare two jars exactly alike, filled with water and containing several snails and some vegetation. Cover one of them completely with black paper, except for a vertical slit one-half inch wide along one side of the jar. Place both jars side by side in a light place, with the slit toward the light. After twenty-four hours remove paper and note quickly the distribution of the snails in both jars. The one jar which is not covered serves as a *control*, that is, it gives us the distribution of snails on the basis of chance. If the snails in the other jar, which is the *experiment*, are not so distributed, then we may conclude that light, the one factor which is different in the two cases, must be responsible for the different distribution of the snails in the two jars. This is the way in which every experiment must be conducted to be convincing. What is the reaction of snails to light? Do they go to the bright light, i.e., are they found collected on the slit, or do they remain in the dark, or are they to be found in weak light? How would this affect their distribution in ponds? The condition which an animal selects when put in a gradient of a particular factor is called its *optimum*. It could similarly be determined which temperature is optimum for the snails by putting them in a trough of water which was warm at one end and very cold at the other; and any other factor can be determined in the same way.

7. **Growth and composition of the shell.**—Observe the parallel lines upon the shell. They are *lines of growth*. The shell is secreted by the mantle at its margin and thus grows continually larger. Do small snails have as many spiral turns as large ones? The distance between successive lines of growth records the rate of growth. Can you find places on the shell where the lines of growth are so close together as to make one deep mark on the shell? How do you explain this? Could you determine in this way how old a snail is? Put an empty shell in hydrochloric acid and observe what happens. What does this indicate as to the composition of the shell? Is the shell entirely soluble in acid? What part is insoluble and why? Does this explain why the shells do not dissolve in the water in which the animal lives? Pond water often becomes acid through the decay of dead organisms in it and the carbonic acid gas produced by the living organisms. Cut out small pieces of shell in living snails, replace in the aquarium and observe whether regeneration occurs.

8. **Reproduction.**—Egg masses of snails will be commonly found in jars where snails are kept. Where are they placed by the snails? Why should the snail not simply drop them to the bottom of the pond in which it lives? Observe the process of development by removing eggs from time to time and studying under the microscope. In what condition are the young snails when they emerge? Are they completely formed?

SUGGESTIONS FOR THE LABORATORY ASSISTANTS

1. **Supplies for students.**—Each student should be provided with a jar or vessel having a tight cover filled with 4 per cent formaldehyde (40 c.c. commercial formaldehyde plus 960 c.c. of water); one or two watch glasses; and a piece of filter paper. The jar should be large enough to receive comfortably the animals used in the course. The student should keep these materials in his locker or elsewhere. Each student will also need a wax-bottomed dissecting pan and finger bowls with glass plates for covers. The dissecting pans are made from low granite-ware pans. The wax used is cerosin (a commercial product obtainable from dealers in laboratory supplies), blackened by lampblack.

2. **Materials for students.**—Three frogs should be allowed for each student. The student should be directed to keep these frogs in the jar of formaldehyde and warned that he will not receive any others. One frog is for the general dissection; a second for the work on general physiology; and a third, injected, for the circulatory system and nervous system. It is always necessary to admonish the student repeatedly not to leave the animals out on the tables and not to allow them to become dry. Of the other animals one specimen should be sufficient.

3. **Section on frog.**—Care should be taken that the slit made in the frog at the end of the first laboratory exercise is through the skin only; if through the muscles also the viscera will protrude through the opening and their relations will be distorted. It is usually advisable to warn against cutting the frog in the median ventral line.

4. **Chemicals required.**—The following solutions and solids are required in the course; they are required in the order named:

a) Dilute acetic acid for the wiping reflex, about 10 per cent, 10 c.c. of pure glacial acetic acid plus 90 c.c. distilled water.

b) Powdered carmine.

c) Artificial gastric juice: 1 to 2 grams of commercial pepsin, obtainable from any drug house, in 100 c.c. of 0.4 per cent hydrochloric acid.

d) 0.4 per cent hydrochloric acid, 4 c.c. of pure hydrochloric acid plus 996 c.c. distilled water.

e) Neutral litmus solution can be made by boiling litmus paper, but should preferably be purchased from dealers.

f) Pancreatic lipase: dissolve one to two grams of commercial pancreatin, obtainable from drug houses, in distilled water. Add enough sodium carbonate to render very slightly alkaline.

g) Starch paste: stir up one gram of starch (corn starch or any commercial product) in a small amount of cold water and pour slowly while stirring into 100 c.c. of boiling distilled water. Allow to boil for a few minutes. Dilute about ten times for using.

h) Glucose solution: 4 or 5 grams of glucose in a liter of distilled water.

i) Fehling's solution consists of two solutions, not to be mixed until ready to be used. First solution, 34.65 grams of copper sulphate made up to 500 c.c. with distilled water. Second solution, 125 grams of potassium hydroxide and 173 grams of Rochelle salt (potassium sodium tartrate) made up to 500 c.c. with distilled water. Keep separately in rubber-stoppered bottles. Just before using mix in equal quantities, and shake until the blue precipitate formed is completely dissolved, yielding a deep blue, clear solution. The assistant should mix the solutions just before the laboratory period and give the students the mixture.

j) Saturated solution of barium hydroxide; should be clear.

k) Physiological salt solution: dissolve 6 to 6½ grams of pure sodium chloride in a liter of distilled water.

l) Aceto-carmine stain (Schneider's): to boiling 45 per cent glacial acetic acid (45 c.c. pure glacial acetic acid plus 55 c.c. distilled water), add powdered carmine until no more will dissolve, and filter.

m) India ink suspension for feeding *Paramecium:* get solid carbon sticks, obtainable in stores dealing in photographic supplies, as it is used for retouching, and rub the stick in tap water until a moderately black suspension is obtained. Do not use ordinary India ink as this may contain toxic substances.

n) Picro-acetic acid: 1 c.c. of pure glacial acetic acid, 99 c.c. of distilled water; saturate the solution with picric acid crystals.

o) Common salt.

A small bottle containing each of these chemicals should be placed on each table in the laboratory to avoid confusion. Needless to say, sugar, starch, and enzyme solutions will not keep and must be made up fresh shortly before being used.

In addition to the above the assistant will require the following:

Ether or chloroform.

Chloral hydrate solution for macerating tissues of the frog. Weigh out 5 grams of chloral hydrate and add enough physiological salt solution to make 100 c.c.

5. Section on general physiology.—All of these experiments have been tried repeatedly and have never been known to fail. No frog should be given out that is not properly pithed, as it will certainly create a disturbance. The assistant should thoroughly familiarize himself with the process of pithing. It should always be done with a blunt instrument, not with a needle. Students should be warned about letting the frog dry up. Commercial pepsin and pancreatin have always been found reliable. A small percentage of human beings have no ptyalin in their saliva.

6. Section on tissues.—This work always proves rather difficult, chiefly because students use too large pieces. Verbal instructions as to the method of procedure will probably be more effective than those written in the outline. Shed epidermis of the frog will be found in the water in which frogs have been kept for a short time. To prepare tissues from the intestine, proceed as follows: Pith a frog. Cut out the small intestine, slit it open, and wash it in physiological salt solution. For columnar epithelium put it into 5 per cent chloral hydrate for twelve to eighteen hours. For smooth muscle, scrape off the mucosa and place the rest, consisting of the muscular coats, in 5 per cent chloral hydrate for twenty-four to forty-eight hours. Connective tissue is always best obtained from a frog which has been preserved in formalin; and striated muscle is usually better from such a source, unless the fresh muscle is carefully handled. Fresh slices of cartilage are infinitely better than any permanent preparations; they are obtained from the ends of the long bones of the frog with a sharp razor. Fresh blood may be obtained by cutting off the toes of a pithed frog and pressing the bleeding ends of the toes against a slide. It is very necessary that the physiological salt solution used for blood be exactly isotonic with the blood, or else the blood cells will become distorted. The myelin sheaths of fresh nerves usually become distorted, producing appearances commonly mistaken for nodes of Ranvier. Students almost always have trouble with the Golgi preparations of nerve cells, often mistaking meaningless deposits of silver on the sections for nerve cells.

7. Section on detailed anatomy.—For the study of the circulatory system, injected frogs are essential. To inject a frog, etherize it, cut off the tip of the ventricle and thrust the cannula of the injection syringe up into the conus arteriosus. All of the arteries and the postcaval vein and its branches are injected in this manner. Veins are best studied in freshly etherized frogs, although they are fairly satisfactory in injected specimens if these are injected shortly before using. We have sometimes let the students work out the veins on freshly etherized frogs and then injected the arteries on the same frogs. This has proved satisfactory. After the frogs are injected a piece of the skull should be removed and the frogs placed in fairly strong formaldehyde, about 8 per cent. This serves not only to harden the injected vessels but also to harden the central nervous system, which can then be studied on the same specimens. The injection mass is made of 800 c.c. glycerin, 1,600 c.c. of water, 3 boxes of cornstarch, 50 c.c. of carbolic acid (melted crystals) or thymol, and enough coloring matter of the color desired (Berlin blue, carmine, chrome yellow, etc.) to give a deep brilliant color to the whole. Stir up before using. In the case of the frog it is not necessary to tie the cannula in the conus nor to tie up the conus after withdrawing the cannula, but simply hold the cannula in with the fingers.

8. Section on embryology.—The frog material used is preserved in formalin. Excellent permanent mounts of halves of early frog embryos can be made as follows: Use slides provided with cells, or make such a cell by sticking hard rubber rings to the slide with balsam. Section the embryos with a sharp razor and mount in the cell in glycerin jelly. The cover glass must be sealed with cement or varnish. To make glycerin jelly, dissolve 6 grams of best gelatin in 42 c.c. of water, and add 50 c.c. of glycerin, and 2 grams of carbolic acid crystals. Warm (not above 75° C.), and stir until homogeneous. It must be liquefied each time used by placing in warm water.

9. Section on Mendel's law.—Culture bottles for *Drosophila* are prepared as follows: Use a wide-mouthed eight-ounce bottle stoppered with cotton. Put a piece of very ripe banana in the bottle and sterilize. Stir up compressed yeast with water to make a paste and drop 2 or 3 c.c. of this on the banana. Put also into the bottle a piece of filter paper to absorb extra moisture. The female flies to be used for a breeding experiment must be isolated from males within a few hours after they emerge from the pupa in order to be certain that they are virgin. The flies which we have been using originally came from the Department of Zoölogy, Columbia University, New York, through the kindness of Professor T. H. Morgan.

10. Section on Protozoa.—Cultures of Protozoa and other micro-organisms may be prepared as follows:

a) Gather a lot of aquatic plants, rotten lily pads, etc., from a pond, pack into a vessel, add just enough water to cover them, and allow to decay. This yields in about a month very good and lasting cultures of *Paramecium*, and other Protozoa.

b) Boil some wheat grains for a few minutes, and put into jars of water in the proportions of about one to two dozen grains to two liters of water. Add a little water, mud, or vegetation from a pond, or if desired to raise a pure culture add a few individuals of the protozoan desired.

c) Boil some hay in a quantity of water for several hours until the water becomes dark brown. Put this brown water into jars with a small quantity of the hay and inoculate with material from a pond or with the Protozoa desired.

These three methods will yield all of the common Protozoa. Such cultures, when they begin to die out, can be rejuvenated within two or three days by adding some crumbs of stale bread.

These methods, particularly the first, will commonly yield small amoebae, but it is difficult to obtain a supply of large amoebae for a large class, and they had better be purchased if possible. Old wheat cultures which have become green often yield large amoebae, as do also cultures of diatoms.

11. Section on *Paramecium*.—To make *Paramecium* stand still, gelatin solutions, jelly made from boiled quince seed, and jelly from *Chondrus* (Irish moss) have been used satisfactorily by many people, particularly the last-named material. We have found that a dilute solution of formaldehyde is remarkably effective for the purpose. Add one drop of formaldehyde to 100 c.c. of water, and test it on *Paramecium*. If too strong continue to dilute until a strength is found which seems to have no effect on the animals immediately but in a few minutes causes them to slow down very gradually. The solution may be used as directed in the manual or a drop of solution may be placed alongside the edge of the cover glass. In the latter case a stronger solution will be required. If the assistant will take the trouble to find out the proper strength by trial in advance, he will find the method to be very successful.

12. **Section on Hydra.**—To obtain *Hydra* collect aquatic plants, particularly soft ones, from clear ponds or bays in sluggish streams. Place the plants in jars, using a large amount of water to a small quantity of the plants. As the animals become noticeable, they may be picked out and placed in a smaller vessel. If they are to be kept for any length of time they must be fed with Entomostraca. To raise Entomostraca place two or three inches of pond mud in a glass vessel, fill with water, and throw in from time to time a few grains of boiled wheat. If Entomostraca do not appear of themselves in such a culture, as they usually will, a few to start it should be obtained from ponds. It is possible by this method to grow the *Hydra* and the Entomostraca in the same vessel, by keeping the fermentation of the wheat to the lowest possible level. The work with the structure of the living animal is always rather difficult for students and may profitably be omitted. Small individuals are best for the purpose.

13. **Section on Planaria.**—*Planaria* occurs in spring-fed pools and in ponds. To collect from springs, hang a piece of meat in the current for an hour or two, whereupon *Planaria*, if present, will attach to the meat and may then be shaken off into a bottle. Collect quantities of plants from ponds and place in pans with a small quantity of water. After a few days, as the plants decay, *Planaria*, if present, will gather at the surface and should be removed. *Planarians* are very intolerant of stagnant water and must be kept in large open pans in which the water is frequently changed. They should be fed every few days with fresh liver. The liver must always be removed within a few hours and the pan and worms thoroughly washed. The digestive tract of *Planaria* will always stand out plainly if the anaesthetized animals are pressed out as described in the manual. However, it can be made to stand out beautifully on the intact animals by feeding them on blood clots shortly before they are to be used. To demonstrate the feeding of *Planaria* use worms which have been starved for a week or two.

14. **Section on the earthworm.**—The dissection of the reproductive system should ordinarily be omitted. Directions are given merely for the sake of completeness.

15. Dealers in supplies.—Living Amoebae can be purchased from A. A. Schaeffer, University of Tennessee, Knoxville, Tennessee. Other Protozoa (sometimes *amoeba* also), *Hydra*, *Planaria*, etc., are sold by Powers & Powers, Station A, Lincoln, Nebraska. All of the preserved material used in the course is obtainable from the Marine Biological Laboratory, Supply Department, Woods Hole, Massachusetts, and from the Chicago Biological Supply House, 5505 Kimbark Ave. Chicago. The last-named firm also sells live frogs and all of the microscopic preparations used in the course.

Abdomen: grasshopper 119, 120; lobster 107
Abdominal appendages: grasshopper 119, 120; lobster 108
Abdominal segment, lobster 107
Absorption, definition of 18
Action of muscle 56
Activities. *See* Behavior
Adrenal gland 11, 37
Air sacs, grasshopper 121
Alimentary canal, definition of 9
Alternation of generations 87, 88
Alternative inheritance 69
Alveoli of lungs 41
Amoeba 72–73
Amoeboid movement 16, 30, 72
Amoeboid organisms 79
Amphipods 125
Anaphase 63
Anemones 88
Animal behavior 135
Animal hemisphere 66
Annelida: in text 80, 95–132; in key 131, 132
Antenna: grasshopper 118; lobster 106, 109, 110, 111
Antennule 106, 107, 109, 110, 111
Anus: frog 3, 40; earthworm 97; grasshopper 120, 123; lobster 107, 114; *Nereis* 96; *Paramecium* 76, 77
Aortic arches, frog 44, 46
Apex of ventricle 9
Appendages: biramous 107; foliaceous 109; grasshopper 117–20; lobster 106–16; typical in lobster 107; uniramous 108
Appendicular: definition of 2; skeleton 51
Aqueduct of Sylvius 48
Arachnids 125
Archenteron 66
Arterial arches 44
Arterial system: frog 44–45; lobster 112, 113, 116
Arteries: frog: carotids 44, coeliaco-mesenteric 45, common iliac 45, cutaneous 5, 44, dorsal aorta 45, pulmonary 44, systemic arch 44; lobster 112, 113, 116
Artery, definition of 9, 30
Arthropoda: in text 81, 106–25; in key 131, 133
Arthropodial membranes 107, 117
Ascaris, mitosis in 62–64
Assimilation, definition of 22, 23
Aster 63
Asterias, development of 65, 66
Atlas 52
Auricle of heart 9, 45, 46
Axial, definition of 2
Axial skeleton 51
Axis cylinder 32
Axone 31, 32

Barnacles 124
Basipod, definition of 107
Battery 82, 84
Behavior: *Amoeba* 73; frog, 15, 19; *Hydra* 83; *Paramecium* 74, 76, 77; *Planaria* 90, 92
Belly of muscle 56
Bilateral symmetry 2, 61
Bile 39; capillaries 33; duct 39, 40
Binomial system 127
Biramous appendage 107
Bladder: gall 10, 39, 40; ligaments of 8, 10; urinary 8, 9, 40, 41

Blastopore 66, 67
Blastostyle 87
Blastula: frog 66; starfish 65
Blood: cells of 30, 31; circulation of, in earthworm 99; circulation of, in frog 20; composition of, in frog 20, 30, 31; corpuscles of 20, 30, 31; function of 20, 25; histology of 30, 31
Blood corpuscles: red 30; white 30, 31
Blood sinuses: grasshopper 120, 121, 122; lobster 112, 113, 115
Blowfly: 20–23; adults 20, 21; egg-laying 20; eggs 21; larvae 21, 22; life-cycle 20–23; pupae 22
Body cavity: earthworm 97, 102; frog 6, 7, 8; origin in embryo 67
Body wall: earthworm 97, 98, 101, 103; frog 5, 7; grasshopper 117, 120, 123; histology of in earthworm 103; lobster 111, 112
Bone, structure of 30
Bones of frog 6, 51–55; kinds of, in skeleton 51; hyoid apparatus 5, 54; limbs 55; lower jaw 53; pectoral girdle 6, 54; pelvic girdle 6, 54; skull 52, 53; sternum 6, 54; upper jaw 53; vertebral column 51, 52
Brachial plexus 48
Brachiopoda in key 130
Brain: earthworm 99, 102; frog 12, 46–48; functions of in frog 49, 50; grasshopper 124; lobster 115; *Planaria* 91
Branchiae, lobster 110
Branchial chamber, lobster 107, 108, 110, 111
Breastbone 6, 54
Brow spot 3, 47
Bryozoa in key 130
Buccal cavity, frog 4, 5
Buccal pouch, earthworm 99
Buds: *Hydra* 86; medusa 87
Bulbus arteriosus 9

Calcareous body, frog 48
Calciferous glands, earthworm 99
Capillary 9, 30
Carapace 106
Carbohydrates: definition of 16; digestion of 18
Cardiac chamber, lobster 114
Cardiac end of stomach, frog 39
Carotid arch 44; arteries 44
Cartilage, structure of 29
Cartilage bone: definition of 51; in skeleton 52
Cell: definition of 24; body of nerve cells 31; division 62–64; membrane 25; structure of typical 25, 26
Cells: chlorogogue 104; of earthworm 103, 104; of frog: blood 30, 31; bone 30, brain 31, cartilage 29, connective tissue 29, epithelial 27, 28, intestine 33, 34, kidney 37, 38, liver 25, 33, motor, of cord 31, 38, muscle 28, 29, nerve 31, reproductive 31, skin 36, 37, spinal cord 38, stomach 35, 36; of *Hydra* 83, 84, 85; of *Planaria* 91; stinging, of *Hydra* 82, 84

frog 33, 34; kidney 37, 38; liver 25, 33; nerve cord, earthworm 104; *Planaria* 91; skin 36, 37; spinal cord 38; stomach 35, 36
Centipedes 125
Central canal, spinal cord 38
Centrosome 63, 64
Centrum of vertebra 51
Cephalization 61
Cephalothorax 106
Cerebellum 47, 50
Cerebral hemispheres 47, 50
Cervical groove 106
Cestodes 94
Chaetae 96
Chaetonotus 80
Chela 109, 112
Chitin 106, 117
Chlorogogue cells 104
Choanae 4
Chordata in key 131
Chordotonal organ 120
Chromatin 25, 63, 70
Chromatophores 16, 20, 36
Chromosomes 63, 64
Cilia: of ciliate Protozoa 79; of epithelium 28; of flatworms 80; of frog 16, 28; of *Paramecium* 75; of *Planaria* 90; of rotifers 80
Ciliary movement, experiment on 16
Ciliate Protozoa 79
Ciliated epithelium 28
Circulation of the blood: in earthworm 99; in frog 20
Circulatory systems: earthworm 98, 99; frog, general 9; frog, special 41–45; function of 20, 23; grasshopper 121–22; lobster 112, 113, 115, 116
Circum-esophageal commissures: earthworm 102; grasshopper 124; lobster 115
Cirri 79, 95, 96
Cisterna magna 11
Class, definition of 128
Classification: definition of 127; exercise on 127–34; of animals 129–34
Cleavage: of *Asterias* egg 65; of frog egg 66; of starfish egg 66
Clitellum 96
Cloaca 40, 41
Clypeus 117
Cnidoblasts 84, 85
Cnidocil 84, 85
Coats of the intestine 33–35
Coelenterata: in text 82–89; in key 129, 132
Coelome: earthworm 97, 98; frog 6–7; grasshopper 120; lobster 112; origin in embryo 67, 71
Coenosarc 87
Colony hydroid 86, 87
Columella 3, 55
Conductivity: definition of 14; experiment on 14, 15
Conjugation, *Paramecium* 78
Connective tissue: definition of 29; kinds of 29, 30
Contractile vacuole: *Amoeba* 72, 73; *Paramecium* 76
Contractility: definition of 15; experiment on 15, 16; of heart 15, 16; of involuntary muscle 15; of voluntary muscle 15
Conus arteriosus 9, 44, 45, 46
Corals 88, 89
Corium of skin 36

Cornea 111
Cornua: of gray matter 38; of hyoid 54
Corpuscles 20, 30, 31
Correlation in nervous system 50
Coxopod, definition of 107
Crabs 125
Cranial nerves 12, 47
Crayfish 106
Crop: earthworm 99; grasshopper 123
Cross-section: comparative 126; earthworm 102, 103, 104; frog 13; *Hydra* 85; *Planaria* 92
Crustacea 124, 125
Cryptoretic organs 9
Ctenophora in key 129
Cultures: methods for Entomostraca 143; *Hydra* 143; *Planaria* 143; Protozoa 142, 143; study of 78–81
Cuticle: earthworm 97, 103; lobster 106; *Paramecium* 75
Cyclops 51
Cystic ducts of liver 40
Cytopharynx 75
Cytoplasm, definition of 25

Daddy longlegs 125
Death, meaning of 1
Defecation, definition of 19
Dendrite 31
Dermal bone, definition of 51
Dermis of skin 36
Development: *Asterias* 65–66; blowfly 20–23; definition of 20, 65; *Drosophila* 20–23; fly 20–23; frog 66–68; general account of 71; starfish 65–66
Diastase: definition of 17; experiment on 18
Diencephalon 47, 48, 50
Digestion: definition of 16, 17; experiments on 17, 18
Digestive apparatus, *Paramecium* 76, 77, 78
Digestive gland, lobster 112, 114
Digestive system: earthworm 99, 100; frog, general 9; frog, special 30–40; functions of 16; grasshopper 123; *Hydra* 82; lobster 114; *Planaria* 91
Dominant character 69
Dorsal aorta 45
Drosophila: breeding experiment 69–70; life cycle 20–23; method of culture 142
Drum membrane 2
Duodenum 10, 39

Ear: frog 3; grasshopper 120
Earthworm 96–104; circulatory system 98; coelome 97; digestive system 99, 100; excretory system 100; external anatomy 96–97; internal anatomy 97–103; microscopic structure 102–4; muscles 102, 103; nerve cord 104; nervous system 101, 103; reproductive system 100, 101
Echinodermata in key 129, 132
Ecology: definition of 135; exercise on 135–38
Ectoderm: embryo 66, 67, 71; fate of, in frog 67; general function of 71; *Hydra* 83, 85; *Planaria* 90, 91
Ectoplasm 72, 75
Ectosarc 75
Egestion, definition of 19
Egg, development of 65–68
Egg guide 120, 122
Egg-laying of: flies 20; grasshopper 120, 122, 123
Eggs: *Ascaris* 62–64; definition of 20; fish 64; flies 20–21; frog 32; sea-urchin 26; starfish 65
Embryology 65–68; *Asterias* 65–66; frog 66–68; starfish 65–66
Endocrinous organs 8
Endophragmal skeleton 115

Endoplasm: *Amoeba* 72; *Paramecium* 75
Endopod, definition of 107
Endosarc 75
Endoskeleton: definition of 51; frog 51–55
Entoderm: embryo 66, 67, 71; fate of, in frog 67; general function of 71; *Hydra* 83, 85; *Planaria* 90
Entomostraca 81, 124
Environment, factors of 135, 136
Enzymes: definition of 17; experiments on 17, 18
Epicranium 117
Epidermis: earthworm 102, 103; frog 27, 36; grasshopper 117; lobster 111; *Planaria* 92
Epimeron 107, 117
Epipharynx 118
Epithelial tissues 27–28
Epithelium: ciliated 28; columnar 27; definition of 27; in earthworm 104; in *Hydra* 85; in intestine of frog 34; in *Planaria* 92; squamous 27; stratified 36
Esophagus: earthworm 99; frog 5, 19, 39; grasshopper 123; lobster 114
Eustachian tube 4
Excretion, definition of 19
Excretory systems: earthworm 100; frog, general 11; frog, special 40, 41; function of 19, 23; grasshopper 123; lobster 115; *Planaria* 91
Exopod, definition of 107
Exoskeleton: definition of 51; grasshopper 117; lobster 106
Eyelids, frog 2
Eyes: frog 2; grasshopper 117; lobster 106, 111; *Nereis* 95; *Planaria* 90, 91; snail, 136

Family, in classification 127
Fascia: definition of 56; dorsal 57; lata 59
Fat body: frog 11; grasshopper 121
Fats: definition of 16; digestion of 17, 18
Feces, definition of 19
Fehling's solution 18, 140
Fertilization: definition of 20, 65; membrane 65
Fibers of connective tissue 29
Fish eggs, mitosis in 64
Fission, *Paramecium* 77, 78
Flagellates 80
Flagellum, definition of 80
Flatworms 80
Flukes 93
Fly: adult anatomy 21; blowfly 20–23; eggs 21; fruit fly 21; larvae 21, 22; life-cycle 21–23; pupae 22
Foliaceous appendage 109
Food ingestion: *Paramecium* 76, 77; *Planaria* 92, 93; snail 136, 137
Food vacuoles: *Amoeba* 73; *Paramecium* 75, 76, 77
Foot, snail 136
Foramen magnum 52
Foramen of Monro 48
Frog (1–61): arterial system 44–45; blood 30, 31; body wall 5–7; brain 46–48; buccal cavity 4–5; circulation of the blood 20; circulatory system: function 20, general anatomy 9, special anatomy 41–46, coelome 6–7; cross-section 13; digestive system: function 16–18, general anatomy 9–10, special anatomy 30–40; excretory system: function 19, general anatomy 5, special anatomy 40–41; external anatomy 1–3; glands of internal secretion 7, 11, 12; heart 45, 46; histology of tissues 24–32; histology of organs: intestine 33, 34, kidney 37, 38, liver 33, skin 36, 37, spinal cord 38, stomach 35, 36; internal

anatomy: general 5–13, special 39–61; mesenteries: general statement 7, 8, of bladder 8, 10, of digestive tract 10, of reproductive system 11; muscles: abdomen 6, 58, function 15–16, general 5–6, hyoid apparatus 58, 59, lower jaw 57, 58, parts of a muscle 56, shank 60, 61, thigh 59, 60, tongue 59; nervous system: function 14, 49, 50, general anatomy 12, special anatomy 46–49; physiology: digestive system 16–18, circulatory system 20, excretory system 19, muscular system 15–16, nervous system 14–15, reproductive system 20–23, respiratory system 18–19, summary 23; reproductive system: general anatomy 4–5, special anatomy 40–41; respiratory system: function 18, 19, general anatomy 10, special anatomy 41; sense organs 12; skeleton: hyoid apparatus 5, 54, limbs 55, lower jaw 53, pectoral girdle 6, 54, pelvic girdle 6, 54, skull 52, 53, sternum 6, 54, upper jaw 53, vertebral column 51, 52; tissues: blood 30, 31, connective 29, 30, epithelial 27, 28, muscular 28, 30, nervous 31–32, reproductive 32; urinogenital system: general anatomy 4, 5, special anatomy 40–41; venous system 41–44
Frons 117
Fruit fly 21–23

Gall bladder 10, 39, 40
Ganglia: earthworm 101, 102, 103; grasshopper 123, 124; lobster 115, 116; spinal 48, 49; sympathetic 12
Ganglion, definition of 49
Gastric: gland, frog 35; juice 17; mill 114; muscles, lobster 111, 114
Gastrolith 114
Gastrovascular cavity: definition of 82; *Hydra* 82, 83; *Planaria* 91
Gastrula: *Asterias* 66; frog 66; general significance 71; starfish 66
Gastrula stage, general 71
Gena 117
Generic name, definition of 127
Genus, definition of 127
Germ layers in embryo 66, 71
Giant fibers 104
Gill: slits of tadpole 68; supports, tadpole 52
Gills, lobster 107, 108, 109, 110
Gizzard: earthworm 99; grasshopper 123
Glands: of frog: adrenal 11, 37, cutaneous 36, gastric 35, mucous 36, 37, poison 37; of internal secretion 8, 11
Glomerulus 38
Glottis 5, 42
Goblet cells 27, 34
Gonads, definition of 10
Gonotheca 87
Gonozooids 87
Grasshopper: abdomen 119–120; appendages 117–20; circulatory system 121–22; digestive system 123; excretory system 123; external anatomy 117–20; head 117–18; internal anatomy 120–24; nervous system 123–24; reproductive system 122–23; sense organs 117, 118, 124; thorax 118–19
Gray matter of cord 31, 38
Green gland 115
Gullet, *Paramecium* 75

Haversian canals 30
Head: earthworm 96; frog 1–5; grasshopper 117–18; lobster 106, 107, 109, 110; *Nereis* 95; *Planaria* 90; snail 136; appendages: grasshopper 117–18, lobster 109–10,

Nereis 95; dominance of 61; of a muscle 56
Heart: frog 9, 45, 46; grasshopper 121, 122; lobster 112, 113
Heart beat 15
Hearts, earthworm 98
Helionoa 79
Hepatic: ducts 40; portal system 42; portal vein 42
Heredity experiment 69–70
Histology: definition of 24; earthworm 102–4; frog, organs 33–58; frog tissues 24–33; *Hydra* 83–85; intestine, frog 33, 34; kidney, frog 37, 38; liver 33; *Planaria* 92; skin, frog 36, 37; spinal cord, frog 38; stomach, frog 35, 36
Horns of gray matter 38
Horseshoe crab 125
Hybrids 69, 70
Hydra: behavior 83; cells of 83–85; cross-section 85; general structure 82; nematocysts 82, 84; reproduction 86; reproductive organs 86
Hydranth 87
Hydroid colony 86–88
Hydrotheca 87
Hyoid apparatus 5, 41, 53, 54; muscles of 58, 59
Hypophysis 48
Hypostome 82

Ileum 39
Inferior lobes 48
Infundibulum 48
Insects 125; in key 131, 133
Insertion of muscle 56
Intercellular substance: definition of 26; in blood 30; in connective tissue 29; in epithelium 27; in muscle 28
Intestine: earthworm 100, 104; frog 9, 10, 39, 40; grasshopper 123; histology of, in earthworm 104; histology of in frog 33–35; lobster 114; *Planaria* 91
Investing bone, definition of 51
Iris 2
Irritability: definition of 14; experiment on 14, 15; of *Amoeba* 73; of *Hydra* 83; of nervous system 14, 15; of *Paramecium* 77; of *Planaria* 90

Jaws: bones of frog 52, 53; fly larva 22; frog 4–5; muscles of frog 57–58
Jellyfish 88

Karyokinesis 62
Key: to classes 131–34; to phyla 129–31
Kidney: frog 11, 37, 40, 41; functions of 19, 23; histology of 37
King crab 125

Labrum: grasshopper 117, 118; lobster 110
Lacuna: bone 30; cartilage 29
Ladder type, nervous system 116
Large intestine, frog 9, 40
Larvae, fly 21–22
Laryngeal chamber 41
Laryngeal prominence 5
Larynx 9, 91
Leeches 105
Legs, parts of: grasshopper 119; lobster 108, 109
Leucocytes 30
Life-cycle fly 20–23
Ligament definition of 7
Ligaments of frog: coronary 9, 10; falciform of liver 8; hepato-gastro-duodenal 10, 39; lateral of bladder 10; median of bladder 8, 10; rectovesical 10; suspensory of liver 8, 10. *See also under* Mesenteries

Limbs of frog: bones of 55; external anatomy of 3; muscles of 59, 60, 61
Linea alba 6
Linin network 25
Linnaeus 127
Lipase: action of 17; definition of 17; experiment on 17
Liver: cells 25; frog 8, 39; histology of 33; lobes of 39; lobster 114
Lobes: of brain 46, 47; of liver 39
Lobster: abdomen 106, 107, 108; abdominal appendages 107; appendages 107–10; circulatory system 112, 113, 115, 116; digestive system 114; excretory system 115; external anatomy 106–11; head 106–7; head appendages 109–10; internal anatomy 111–16; muscles 111–12; nervous system 115–16; reproductive system 113–14; respiratory system 110–11; segmentation 106–16; sense organs 111; thoracic appendages 107–8, 109, 110, 111; thorax 106, 107, 110, 111
Lungs 9, 41
Lymphocytes 31
Lymph 5: sac subvertebral 11; sacs, frog 5, 11, 13; spaces, frog 5, 11

Malpighian bodies of kidney 37; tubules 123
Mandible: frog 53, 57; grasshopper 118; lobster 109, 110
Mandibular arch, frog: bones of 53; muscles of 57
Matrix: of bone 50; of cartilage 29; of connective tissue 29
Maxilla: grasshopper 118; lobster 109, 110
Maxillary: arch, bones of 53; teeth 4, 53
Maxillipeds 108, 109
Meckel's cartilage 53
Medulla oblongata 47, 50
Medullary folds 67; sheath 32
Medusa 87, 88
Medusa buds 87
Medusae 87, 88
Membrane bone 51, 52
Mendel 70
Mendel's law 69, 70
Mesenteries, earthworm 97
Mesenteries, frog 7, 8, 10, 11; dorsal 8, 10; dorsal of liver 10; mesentery of intestine 10; meso-esophageum 10; mesogaster 10; mesorchium 11, 40; mesorectum 11; mesotubarium 11; mesovarium 11, 41; ventral 8, 10. *See also under* Ligaments
Mesentery, definition of 7
Mesoderm: embryo 66, 67, 71; fate of, in frog 67; general significance of 71; *Planaria* 90, 92
Mesogloea, 83, 85
Mesothorax 118, 119
Metabolism, definition of 23
Metagenesis 87, 88
Metamere, definition of 95
Metamorphosis, definition of 22
Metaphase 63
Metathorax 118, 119
Millipedes 125
Mites 125
Mitosis 62–64
Mitotic figure 63
Mollusca in key 130, 133
Molluscoidea in key 130
Morphology, definition of 14
Mouth: cavity: earthworm 99, frog 44; earthworm 96, frog 4; grasshopper 118; *Hydra* 82; lobster 107, 110; *Nereis* 95; *Planaria* 75; *Planaria* 91; parts: grasshopper 117, 118; lobster 107, 109, 110
Mucus (33, 103, 90): cells: earthworm 103, *Planaria* 90, 92; glands: earthworm 103, frog 36, *Planaria* 92; thread of snail 136

Muscle: cells, structure of 28, 29; contractility of 15; function of 15, 16; heart 15, 16; involuntary 15, 28; kinds of 15, 28, 29; microscopic structure of 28, 29; parts of 56; smooth 28; striated 28, 29; typical 56; voluntary 28, 29
Muscles of: abdomen, frog 58; body wall earthworm 102, 103; body wall frog 5–6; chela 112; earthworm 102, 103; frog 5, 6, 15, 56–61; grasshopper 121; *Hydra* 85; hyoid apparatus 58; lobster 111, 112; lower jaw frog 57, 58; *Planaria* 92; shank frog 60; thigh frog 59; tongue frog 58, 59; trunk frog 57, 58
Muscular: system: earthworm 102, 103, frog 5, 6, 15, 56–61, function of 15; grasshopper 121, *Hydra* 85, lobster 111, 112, *Planaria* 92; tissue, structure of 28, 29
Myelin 32

Naids 80
Nares: external 2; internal 4; method of closure 58; respiratory movements of 18, 19
Nemathelminthes 80; in key 130
Nematocysts 82, 84, 85
Nematodes 80
Nemertines in key 130
Nephridia: earthworm 98, 100, 102; lobster 115
Nephridiopore 100
Nephrostome 100
Nereis 95, 96
Nerve: conductivity in 14, 15; definition of 33; irritability of 14, 15; roots of 51; sciatic 15; stimulation of 15; structure of 32
Nerve cells: brain 31; earthworm 99, 102, 104; frog 31, 33 38; lobster 111; motor of cord 31, 32; *Planaria* 90, 91; spinal cord frog 38; structure of 31, 32
Nerve cords: earthworm 99, 102, 104; grasshopper 123, 124; lobster 115, 116; *Planaria* 92
Nerve plexus, frog 48
Nerves: cranial, frog 12, 47; roots of 51; spinal, frog 12, 48, 49
Nervous system: central, frog 12, 46–50; earthworm 100, 102; frog 12, 46–50; function of 14, 46–50; grasshopper 123, 124; *Hydra* 86; lobster 115, 116; origin of, in embryo 67; peripheral, frog 12, 47–49; *Planaria* 91, 92; sympathetic, frog 12
Neural: arches 12, 51; canal 12, 47, 51; fold 67; spine 51
Neurilemma 32
Neuroglia 38
Nictitating membrane 4
Nostrils 2
Notochord 67
Notum 117
Nuclear membrane 25
Nucleolus 26, 64
Nucleus: *Amoeba* 73; definition of 24; in mitosis 62–64; *Paramecium* 76; structure of 25, 26
Nutritive zooids 87

Obelia 86–88
Ocelli 117
Occipital condyles 53
Olfactory lobes 46, 50; nerve 47; sense: frog 50; grasshopper 124; lobster 111; *Planaria* 91, 92
Ommatidia 111, 117
One-celled animals 72–80
Ontogeny, definition of 20
Optical section, definition of 65
Optic chiasma 48; ganglion, lobster 111; lobes 47, 50; nerves, frog 47
Oral groove 74
Orbit 2, 52
Order in classification 128

Organ, definition of 8
Origin of a muscle 56
Ossicles 114
Ostia of heart: grasshopper 122; lobster 113
Ostium of oviduct 41
Ova: definition of 20; of fly 21; of frog 10, 32; of sea-urchin 26; of starfish 65; structure of 26
Ovaries: earthworm 101; frog 9, 10, 41; function of 20; grasshopper 121; Hydra 86; lobster 114
Oviducta: earthworm 101; frog 11, 41; grasshopper 122; lobster 114; male frog 11, 40; ostium of 41
Ovipositor 120, 122
Oxidation, definition of 19, 23

Palps: grasshopper 118; Nereis 95
Pancreas 9, 39, 40
Pancreatic ducts 40
Paramecium 73–78; avoiding reaction 77; behavior 74, 76, 77; conjugation 78; digestive apparatus 75, 76; experiments on 76, 77; fission 77–78; food ingestion 76, 77; reaction to chemicals 77; reproduction 77–78; structure 74–76
Parapodium 95, 96, 105
Parenchyma 92
Pectoral girdle: bones of 6, 54; definition of 6: muscles of 58
Pelvic girdle: bones of 6, 54, 55; definition of 6; muscles of 58, 59, 60
Penis 120, 122
Pepsin 18
Pereiopods 108, 109
Pericardial cavity 7; sac 7
Pericardium 7
Perisarc 87
Peristalsis, definition of 15
Peristomium 95, 96
Peritoneum: definition of 7; earthworm 98; frog 6, 7, 8, 11; origin of, in embryo 67; parietal 7, 98; visceral 7, 98
Pharyngeal chamber 91, 92
Pharynx: earthworm 99; frog 39; Nereis 95; Planaria 91, 92
Phyla, key to 129, 131
Phylum, definition of 72, 128
Physiology: definition of 72; of circulatory system 20; of digestive system 16–18; of frog 14–23; of excretory system 19; of muscular system 15–16; of nervous system 14, 15, 49–50; of reproductive system 20–23; of respiratory system 18, 19; of skin 19; summary 23
Pia mater 38, 46
Pigment cells: liver 33; skin 16, 36, 20; granules: liver 33; Planaria 90
Pill bugs 125
Pineal body 3, 47
Pitching, method of 1
Planaria: behavior 90; cross-section 92; digestive system 91; food ingestion 92, 93; histology 92; regeneration 93; sense organs 90, 91; structure 90–93
Plasma, definition of 20, 30
Plasmosome 26
Platyhelminthes: in key 130, 132; in text 80, 90–94
Podical plates 120
Polychaetes in key 132; in text 105
Polymorphism 87–88
Polyp 87
Pond snail 135–138. See under Snail
Porifera in key 129
Portal system: definition of 43; hepatic portal 43; renal portal 43
Prehallux 3
Proboscis, Planaria 91
Pronotum 118, 119
Prophase 63
Prostomium 95, 96
Protease: action of 17; definition of 17; experiment with 17

Proteins: definition of 16; digestion of 17
Prothorax 118, 119
Protoplasm, definition of 24
Protopod, definition of 117
Protozoa: ciliate 79; flagellate 80; definition of 72; in key 129, 131; in text 72–80
Pseudopodia 72
Pseudothyroid 12
Ptyalin 18
Pulmonary artery 44; sac, snail 137; vein 43
Pulse 20
Pupae, flies 22
Pupil 12
Pyloric chamber 114; division, stomach 39
Pylorus 39

Radula 136, 137
Ramus communicans 12, 49
Reaction, definition of 14
Reaction time: definition of 14; of involuntary muscle 15; of nerve 14, 15; of voluntary muscle 15
Reactions of animals. See Behavior
Recessive character 69
Rectum: frog 40; grasshopper 123; lobster 114
Reflex: definition of 14, 50; in frog 14, 15, 50; wiping 14, 15
Regeneration, Planaria 93
Renal portal system 43
Reproduction, method of: earthworm 96; flies 20–23; grasshopper 120, 122, 123; Hydra 86; Paramecium 77, 78; snail 138
Reproductive cells 32; systems: earthworm 100–101, frog, general 10, frog, special 40, 41, function of 20–23, grasshopper 120, 122–23, Hydra 86, lobster 113, 114
Respiration: definition of 18; external 18; frog 19; function of 18, 19, 23; insect larvae 21; internal 19; muscles of 58, 59; grasshopper 121; snail 137; through skin of frog 19
Respiratory function of the skin 19; movements of frog 18, 19; sac 137; system: fly larvae 21; fly pupae 22, frog, general 9, frog, special 41, function of 18, 19, grasshopper 121, insects 22, 121, lobster 110, 111
Response, definition of 14
Resting cell 62
Retroperitoneal, definition of 11
Rhabdites 90, 92
Ribs 6
Root cap plants 64
Root tips, mitosis in 64
Roots of spinal nerves 47, 49
Rostrum 106
Rotifers: in key 130; in text 80
Roundworms: in key 130; in text 80
Rugae, stomach 35, 39

Sacral vertebra 52
Sagittal axis 2; plane 2, 61
Saliva, action of 18
Sarcolemma 28
Sarcostyles 29
Sciatic nerve 15; plexus 48
Sclerite 117
Scorpions 125
Sea fans 89; feathers 89
Segmentation: cavity 65; earthworm, coelome 97, 98; earthworm, body 96; of the egg 65–66; embryo 68; fly 21–22; frog, adult 61; frog, tadpole 68; general discussion of 61, 68, 71; grasshopper 117; insects 21, 22, 117; Nereis 95, 96; lobster 106, 116; tadpole 68
Segments: definition of 21, 61, 68; embryo 68; earthworm 96; fly 21–22; frog 61; grasshopper 117–20, 124; insects 22–23; lobster 106, 107–110, 113, 116; Nereis 95, 96; tadpole 68

Seminal funnel 101; receptacles 98, 100, 101; vesicles 98, 100, 101
Sense centers, brain 49, 50
Sense organs: earthworm 96; frog 12; grasshopper 117, 118, 124; lobster 111, 107, 108; Nereis 95; Planaria 90, 91
Septa 98
Serosa: definition of 7; of intestine 34, 35
Setae: earthworm 97, 102, 103; Nereis 96
Shank 3; bones of 55; muscles of 60, 61
Shell, snail 136, 138
Shrimps 125
Sinus, blood: grasshopper 120, 121, 122; lobster 112, 113; pericardial 112; sternal 113, 115
Sinus venosus 9, 45
Siphonophora: in key 132; in text 88
Skeleton: breastbone 54; definition of 5, 51; endophragmal 115; endoskeleton, definition of 51; exoskeleton, definition of 51; frog, general 5, 6; frog, special 51–55; grasshopper 117, 123; hyoid apparatus 53, 54; jaws, frog 52, 53; limbs, frog 55; lobster 106, 115; origin of 51, 67; pectoral girdle, frog 54; pelvic girdle 54, 55; skull 51, 52–53; sternum 54; vertebra, parts of 51, 52; vertebral column 51, 52; visceral skeleton 52–54
Skin: external features of 3; histology of 36–37; glands of 36–37; in respiration 19; in skeleton 51
Skull: bones of 51–53; definition of 6; formation of 52
Small intestine 9, 39; histology of 36, 33–35
Snail: 135–138; behavior 135–36; food ingestion 136–37; radula 136–37; reaction to drying 137; reaction to lack of oxygen 137; reaction to light 137–38; reproduction 138; respiration 137; shell 136–38; structure 135–36
Somite 95
Sow bugs 125
Species, definition of 127
Specific name, definition of 127
Spermatozoa: definition of 20; earthworm 101; frog 32; structure of 32
Spiders 125
Spinal cord, frog: cells of 31, 38; function of 14, 15, 49; gross structure 14, 47, 48, 49; histology of 38
Spinal nerves: frog 12, 47, 48, 49; roots of 49
Spindle 63
Spiracles: fly 21–22; grasshopper 119, 120, 121
Spireme 63
Spleen 10, 11
Starch test 18
Starfish, development of 63–66
Statocyst 111
Stentor 79
Sternum: frog 6, 54; lobster 107; grasshopper 117, 119, 120
Stigma 110, 120
Stinging cells 82, 84
Stomach: frog 9, 39; glands of 35; grasshopper 123; histology of 35, 36; lobster 111, 114
Styles 120
Sub-esophageal ganglion: earthworm 100; grasshopper 123; lobster 115
Sugar test 18
Sulcus marginalis 4
Supra-esophageal ganglion: earthworm 102; grasshopper 124; lobster 115
Suture 117
Swimmerets 108
Syncytium, definition of 29
Systemic arch 44, 45
Systems: comparison of 126; of frog 8

Tail fan, lobster 108
Tapeworms: in key 132; in text 94
Taxonomy, definition of 127
Teeth: frog 4; lobster 114
Telophase 63
Telson 107
Temperature, effect of, on activity 15
Tendon 56; of Achilles 56
Tentacles: *Hydra* 82, 83, 84; *Nereis* 95; snail 136
Tergum 107, 117
Testes: earthworm 101; frog 11, 40; function of 20; grasshopper 122; *Hydra* 86; lobster 114
Thalamencephalon 47
Thigh, 3; bones of 55; muscles of 59, 60
Thoracic appendages: grasshopper 119; lobster 108, 109
Thorax: grasshopper 117, 118, 119; lobster 107, 110, 111
Thyroid gland 12
Tissue, definition of 26
Tissues: connective 29-30; epithelial 27, 28; frog 26-33; muscular 28; nervous 31
Tongue 4; muscles of 58, 59
Tract, definition of 8
Tracts of frog 8
Tracheae. *See* Tracheal tubes
Tracheal tubes: fly larvae 21, 22; grasshopper 119, 121

Transverse processes 6, 51
Trematodes: in key 132; in text 93
Trichocysts 75, 77
Triploblastic 71, 92
Trochelminthes: in key 130; in text 80
Truncus arteriosus 44, 46
Trunk, muscles of 57
Turbellaria: in key 132; in text 93
Tympanic membrane 2; ring 57
Typhlosole 100, 103

Undulating membrane 76
Uniramous appendage 108
Unit character 70
Ureter 11, 40
Urinary bladder, frog 8, 9, 40, 41, 10; ligaments of 8, 10
Urinogenital system, frog: definition of 11; general 10, 11; special 40, 41
Uropods 108
Urostyle 6, 51
Uterus 41

Vacuoles: contractile 72, 73, 76; food 73, 76, 77
Variety, definition of 127
Vasa efferentia 40
Vas deferens: earthworm 101; grasshopper 122; lobster 114
Vegetative hemisphere 66
Vein, definition of 9, 30

Veins: of frog 5, 6, 42-43, abdominal 6, 43, 8, hepatic portal 43, musculo-cutaneous 5, 42, pulmonary 43, postcaval 42, precaval 42, renal portal 43, systemic 42; of wings 119
Venous system, frog 42-43
Ventricle of heart 9, 45, 46
Ventricles of brain 46, 47, 48
Vertebra, structure of 51
Vertebral column 6, 51, 52
Vertebrata in key 131, 133, 134
Vertebrates, definition of 72
Vertex 117
Viscera, definition of 5
Visceral skeleton 52, 53, 54
Vocal cords 41; sac 5
Vomerine teeth 4
Vorticella 79

Water bears 81; fleas 81; mites 81
Whitefish eggs, mitosis in 64
White matter of cord 31, 38
Wings 119
Wolffian duct, 11

Yolk 66
Yolk plug 66, 67

Zooids 87-89
Zygapophyses 52

Milton Keynes UK
Ingram Content Group UK Ltd.
UKHW020211300524
R3595700001B/R35957PG443077UKX00007B/1

9 781018 372884